城市住区

中国美术学院&法国巴黎美术学院
城市设计联合教学实录

卓旻　著

中国美术学院出版社

目 录

序

不同的观点
3　现代主义
9　功能混合
15　类型问题
21　封闭社区
27　拥挤文化
33　山水城市

时代的案例
41　理想公社
43　马赛公寓
45　老区新楼
47　类型研究
49　公共组屋
51　生长社区

两块 场地
55　之江地块
95　运河地块

Contents

Preface

Diverged Thoughts
Functionalism   3
Generators of Diversity   9
Typological Study   15
Gated Community   21
Conjestion Culture   27
Landscape City   33

Case Studies of Ages
Utopian Commune   41
Unité d'Habitation ait Marseilles   43
New Complex Set in Historic Neighborhood   45
Typological Residence   47
Public Housing   49
Ever-growing Community   51

Two Pieces of Land
Zhijiang River Field   55
Canal Field   95

序
PREFACE

中国城市发展到今天，最大的共性特征其实不在于那高耸林立的天际线，而就在我们所处的那一个个封闭式的居住小区。中国式城市住区的面貌是中国各方面制度叠合的最终结果，也是当今中国城市诸多问题的主要源起。

回顾我们居住小区的发展，新中国成立后的第一次城市建设热潮中的一大特点就是随着工业发展而兴起的众多单位大院，大院内一应俱全的设施往往让大院外的人们羡慕不已。而在改革开放之后，为了推动城市居民的生活水准，住房改革引发的对于房地产的大量需求开始催生中国的房地产业发展。中国的土地出让制度、规划管理制度、地方财政制度是这几十年来封闭式住区盛行的根本原因。将大片土地归整到一起进行征地、拆迁、平整、开发，对于相关从业技术人员不足的中国来讲，是提高规划效率和开发效率的不二法门。中国的房地产开发商时常可以拿到占地几十公顷的大块土地进行开发建设，这在英美这类土地私有化的国家是难以想象的。同时为了节省财政开支，地方政府在出让土地时，通常要求"谁开发谁配套"的模式。开发商在负责房产开发的同时，还要负责相关公共建筑和公园道路等的建设。而这些公共设施的建设费用最终当然是摊派到小区的业主头上，既然如此，那么从开发商到业主都有一致的动力来实现这些公共设施在使用上的排他性。不仅如此，公安部门也从安全考虑出发推动并要求居住小区进行封闭，以降低罪案发生率，当然这也完全符合小区业主对于自身财产安全保障的需求。

城市生活中公共性与私密性是一对平衡的关系，当私密性被过分放大时，公共性势必要做出牺牲。大尺度的封闭式居住小区将城市变成一个个孤岛的集合，每一个孤岛在获得其私密性的同时，而牺牲掉了外面的城市公共空间。在这样的城市里，交通问题是没有可能解决的，因为尽管看似交通要道不断拓宽，城市中还增添了高架道路，但实际的道路密度却因为社区的封闭而大大低于国外同类城市。街道少了，沿街的店铺自然也少，而租金却会高启，街道商业势必难以发展起来，需要商业带动的城市生活也就越来越无趣。

政府部门出于财政收入的要求，往往希望能从土地买卖中获得最大的利益。这其中最简单的方式莫过于提高容积率了，容积率高的地块允许建造更多的住宅，土地价值自然就上去了。同时为了僵硬地追求绿化密度（其实质是空地密度），规划部门又定下了低建筑密度的要求。这样就自然形成了封闭小区高楼林立的外貌，从而也决定了一个城市的风貌特征。

"要么建筑，要么革命"，这是柯布西耶为西方社会进入工业时代以来的社会矛盾所开的建筑药方。换言之，就是为了避免革命的血腥，唯一的办法就是为大众建造更多的住宅以改善他们的生活条件（从而降低他们对于革命的热情）。所以，居住问题一直是柯布西耶的现代主义城市规划中的核心问题。而我们也都熟知他所设计的集合住宅——马赛公寓。法国社会的高福利制度使得法国建筑师一直延续着对于城市住区的关注。而同时，法国人又面临着他们伟大的城市历史，现代的居住社区如何与传统街区共存一直是法国建筑师的难题。

城市住区不仅是城市设计学科当中的一个核心问题，似乎也是中法两国建筑师之间最具有共通性的问题。由是，当接手中法教学的国际交流任务后，我特意选取了在这中法两个不同社会制度中但却有相同关注度的居住社区为题目，进行了延续三年的联合教学实践。希望不仅让学生们获得设计上的锻炼，也能为中国城市住区的发展提供一些不同的关注角度。自2012年始，每年春季学期，由我选取位于杭州的具有一定当代城市化发展特征的场地，拟定题目和关注的角度，邀请法国巴黎美术学院的马拉盖国立高等建筑设计学院 [École Nationale Supérieure d'Architecture de Paris-Malaquais] 的师生来到中国美术学院进行合作教学。我们的学生和法国的学生混合编组，共同进行实地调研，共同进行概念设计，然后在中期评图结束后，各自回到自己学校进行设计深化，从而完成各自的设计方案。在合作教学过程中，中法双方的同学需要能了解中国当下这一前所未有的城市化进程的大背景，而最重要的是要求他们从场地的相关性调研中挖掘出独特的角度来关注中国的城市住区问题，从而引发他们的设计。这些角度涵盖历史传统、环境地理、社会伦理、规划制度、基础设施、自然资源等等，其目的是丰富学生们独立思考的设计能力，不希望他们陷入到一种单一的思维模式。两方同学都从这一合作教学模式中获益良多，开拓了各自的国际视野和思考问题的角度。

因工作繁忙，直到今年才将这几年前的教学成果进行了整理，匆匆付梓。本书分为三大部分。第一部分介绍现代主义以来学者们对于城市和住区的不同观点，学术观点无所谓对错，都是在一定历史时期针对某个特定问题提出来的，有必要让学生了解相关学术思想的发展，避免以偏概全。第二部分介绍了不同历史时期的一些具有代表性的城市住区的案例，虽然案例不多，但大概能对城市住区的不同发展类型有一个直观的了解。第三部分是将国际交流的教学成果以不同的视角进行展现，课程教学对象是刚进入专业学习的三年级同学，所以方案只能停留在概念阶段，但至少应该能给相关从业人员提供一些看问题的不同切入点，也算是对参加课程实践的同学有所交待。

也在此通过本书谨以记录在中国城市化的大背景下，我们作为设计人员在面对当下城市当中的封闭住区并不是一味的无奈，也曾经对此进行过一些挣扎。最后，通过本书的出版，希望能给建筑专业的师生在进行住区设计时提供一个教学参考，也希望有更多人能关注中国的居住社区问题和城市面貌问题。

卓旻
2016年11月2日

不同的观点

DIVERGED THOUGHTS

现代主义

A Contemporary City of Three Million Inhabitants

…
We must have some rule of conduct.
We must have fundamental principles for modern town planning.

SITE
A level site is the ideal site. In all those places where traffic becomes over-intensified the level site gives a chance of a normal solution to the problem. Where there is less traffic, differences in level matter less.
The river flows far away from the city. The river is a kind of liquid railway, a goods station and a sorting house. In a decent house the servants' stairs do not go through the drawing-room - even if the maid is charming (or if the little boats delight the loiterer leaning on a bridge).
POPULATION
This consists of the citizens proper; of suburban dwellers; and of those of a mixed kind.
(a) Citizens are of the city: those who work and live in it.
(b) Suburban dwellers are those who work in the outer industrial zone and who do not come into the city: they live in garden cities.
(c) The mixed sort are those who work in the business parts of the city but bring up their families in garden cities.
To classify these divisions (and so make possible the trans-mutation of these recognized types) is to attack the most important problem in town planning, for such a classification would define the areas to be allotted to these three sections and the delimitation of their boundaries. This would enable us to formulate and resolve the following problems:
1. The City, as a business and residential centre.
2. The Industrial City in relation to the Garden Cities (i.e. the question of transport).
3. The Garden Cities and the daily transport of the workers.
Our first requirement will be an organ that is compact, rapid, lively and concentrated: this is the City with its well-organized centre. Our second requirement will be another organ, supple, extensive and elastic; this is the

三百万居民的现代城市

……
必须有行为准则
必须有现代城市规划的基本原则

场地

平坦之处是理想之场地。各处的交通流量都更加紧凑，平坦的地势能够提供正常的解决办法。平坦之处的交通流量降低了，令人困扰的意外也将会减少。

车流远离城市。车流成了一种流动的铁路，即是货运站，也是调配站。在一栋精心设计的房间里，仆人用的侧梯是不会穿过客厅的——即使说布列塔尼的女仆们喜欢卖弄风情（或即使说海上的游艇喜欢炫耀和取悦桥上弯着身子看热闹的人们）。

人口

城市人口，郊区人口以及二者的混合人口。
a）城市人口，生活重心在城市里并且居住于城市之中者。
b）郊区人口，生活于郊区的工业区且不住在城市之中者；他们住在花园新城里。
c）二者的混合人口，工作于城市的商业区之内，但居住在花园新城。

将a、b、c各项予以分类（从而使各种公认类型的转化成为可能），旨在回应城市规划最重要的问题，因为这将涉及三个部分的土地区划，确定其范围，这样才能提出并解决下述问题：
1. 城市，作为商业和居住的中心；

2. 工业城与花园新城（交通问题）；
3. 花园新城与工人的日常交通。

辨别出一个紧凑、快速、敏捷、集中的机制：城市（妥善组织的中心）。另一个则是流畅、广阔、富有弹性的机制：花园新城（外围地区）。

在这两种机制中，应注意到立法确定保护区与拓展区，如允许拓展的保护区、森林和草原区、航空预备区等存在的绝对必要性。

密度

一个城市的人口密度越高，其行程就越短。结论是：提高城市中心和商业中心的密度。

肺

现代化的工作变得越来越紧张，总是更危险地刺激着我们的神经系统。现代工作需要的是清静，以及没有污染而有益健康的空气。

为了提高密度，目前的城市牺牲掉了城市之肺与植被。

新的城市应该在提高其密度的同时，也大量地增加植被的面积。

增加植被的面积并减少行程的距离。必须兴建向垂直方向发展的城市中心。

城市中的住房不能再建造那些充斥着混乱、弥漫着尘埃的"甬道"之旁以及阴暗的中庭之中。

城市中的住房应当建成没有中庭、远离道路、窗户面向广阔的公园：锯齿状住宅社区和封闭式住宅社区。

道路

目前的道路是一种古老的"牛皮地板"式的铺设，下面挖出了几条地铁。

现代化的道路应当是一种新的机制，是那种长条形工厂，通风良好且充满复杂精细装置（管道系统）的集散地。将城市的管道系统埋在地下完全是违反经济、安全与理性的做法。管道系统应当随处都能进入。这座长条形工厂的各层都有各种特别的用途。将要实现的这座工厂，就像我们习惯于在两侧兴建的房屋，以及横跨山谷或溪流之上的桥梁一样。

现代化的道路必须是土木工程师的杰作而不

Garden City on the periphery.
Lying between these two organs, we must require the legal establishment of that absolute necessity, a protective zone which allows of extension, a reserved tone of woods and fields, a fresh-air reserve.

DENSITY OF POPULATION
The more dense the population of a city is the less are the distances that have to be covered. The moral, therefore, is that we must increase the density of the centres of our cities, where business affairs are carried on.

LUNGS
Work in our modern world becomes more intensified day by day, and its demands affect our nervous system in a way that grows more and more dangerous. Modern toil demands quiet and fresh air, not stale air.

The towns of to-day can only increase in density at the expense of the open spaces which are the lungs of a city.

We must increase the open spaces and diminish the distance to be covered. Therefore the centre of the city must be constructed vertically.

The city's residential quarters must no longer be built along "corridor-streets," full of noise and dust and deprived of light.

It is a simple matter to build urban dwellings away from the streets, without small internal courtyards and with the windows looking on to large parks; and this whether our housing schemes are of the type with "set-backs" or built on the "cellular" principle.

THE STREET
The street of to-day is still the old bare ground which has been paved over, and under which a few tube railways have been run.

The modern street in the true sense of the word is a new type of organism, a sort of stretched-out workshop, a home for many complicated and delicate organs, such as gas, water and electric mains. It is contrary to all economy, to all security, and to all sense to bury these important service mains. They ought to be accessible throughout their length. The various storeys of this stretched-out workshop will each have their own particular functions. If this type of street, which I have called a "workshop," is to be realized, it becomes as much a matter of construction as are the houses with which it is customary to flank it, and the bridges which carry it over valleys and across rivers.

The modern street should be a masterpiece of civil engineering and no longer a job for navvies.

The "corridor-street" should be tolerated no longer, for it poisons the houses that border it and leads to the construction of small internal courts or "wells."

TRAFFIC
Traffic can be classified more easily than other things.
To-day traffic is not classified-it is like dynamite flung at hazard into the street, killing pedestrians. Even so, traffic does not fulfill it: function. This sacrifice of the pedestrian leads nowhere.

If we classify traffic we get:
(a) Heavy goods traffic.

(b) Lighter goods traffic, i.e. vans, etc., which make short journeys in all directions.
(c) Fast traffic, which covers a large section of the town.
Three kinds of roads are needed, and in super-imposed storeys:
(a) Below-ground there would be the street for heavy traffic. This storey of the houses would consist merely of concrete piles, and between them large open spaces which would form a sort of clearing-house where heavy goods traffic could load and unload.
(b) At the ground floor level of the buildings there would be the complicated and delicate network of the ordinary streets taking traffic in every desired direction.
(c) Running north and south, and east and west, and forming the two great axes of the city, there would be great arterial roads for fast one-way traffic built on immense reinforced concrete bridges 120 to 180 yards in width and approached every half-mile or so by subsidiary roads from ground level. These arterial roads could therefore be joined at any given point, so that even at the highest speeds the town can be traversed and the suburbs reached without having to negotiate any cross-roads.
The number of existing streets would be diminished by two-thirds. The number of crossings depends directly on the number of streets; and cross-roads are an enemy to traffic. The number of existing streets was fixed at a remote epoch in history. The perpetuation of the boundaries of properties has, almost without exception, preserved even the faintest tracks and footpaths of the old village and made streets of them, and sometimes even an avenue (see Chapter I: "The Pack-Donkey's Way and Man's Way ").
The result is that we have cross-roads every fifty yards, even every twenty yards or ten yards. And this leads to the ridiculous traffic congestion we all know so well.
The distance between two bus stops or two tube stations gives us the necessary unit for the distance between streets, though this unit is conditional on the speed of vehicles and the walking capacity of pedestrians. So an average measure of about 400 yards would give the normal separation between streets, and make a standard for urban distances. My city is conceived on the gridiron system with streets every 400 yards, though occasionally these distances are subdivided to give streets every 200 yards.
This triple system of superimposed levels answers every need of motor traffic (lorries, private cars, taxis, buses) because it provides for rapid and mobile transit. Traffic running on fixed rails is only justified if it is in the form of a convoy carrying an immense load; it then becomes a sort of extension of the underground system or of trains dealing with suburban traffic. The tramway has no right to exist in the heart of the modem city.
If the city thus consists of plots about 400 yards square, this will give us sections of about 40 acres in area, and the density of population will vary from 50,000 down to 6,000, according as the "lots" are developed for business or for residential purposes. The

再只是一种挖土工程而已。

我们不能再允许甬道式道路的存在，因为它们腐蚀了位于其上的住宅并导致了封闭式天井的存在。

交通

交通的分类——优于其他任何事情。

目前的交通并无分类——就像一颗丢在甬道中人群里的炸弹。行人全给炸死掉了。而且因为它的缘故，交通不能再运行。行人的牺牲毫无意义。

对交通进行分类：
a）载重卡车；
b）低速卡车（向各个方向短途行驶）；
c）高速汽车（横越大半个城市）。

需要三种类型的道路，分层布置：
a）地下层，载重卡车。该层由房屋架空柱所构成，其间形成宽大的开阔空间，载重卡车在此装卸货物，偶然构成了房屋的码头。
b）建筑物的一楼平面，与普通道路形成复杂而精巧的运输系统，引导各方向的交通至其目的地。
c）横贯东西、纵贯南北的两条城市轴线，置快速单向车道，建于40～60米宽的混凝土高架桥上，每隔800～1200米有坡道相连至普通道路平面上。我们能够在其中的任一坡道处进入快速干道并驶过城市、抵达郊区，以最快的速度，并不需要忍受任何十字路口的阻扰。

现状道路的数量必须减少三分之二。道路十字路口的数量也紧随道路的数量而变化；目前道路数量的情况十分糟糕。道路十字路口是交通的敌人。现状道路的数量源自于久远以前的历史。由于对土地产权的保护，几乎毫无例外地仅仅保持了早期甬道的格局并升格为城市道路，甚至林荫大道有时也是如此（见第一章《驴行之道与人行之道》）。像这样的道路每隔50米、20米甚至10米便要相交！然而这正是荒谬的交通拥堵的导火索啊！

两个地铁站或公交车站的间距提供了道路十字路口间距的有效模距，而模距本身则受到行车速度与行人可接受步行距离的限制。通常400米的均值提供了城市距离的基本标准和道路的一般间距。我的城市正是依照道路间距400米、偶尔再进一步细分为200米的规则所做的区划。

这种由三部分重叠起来所组成的道路系统，

可以适应各种高速与低速行驶的汽车交通运输（卡车、出租汽车或私人轿车、公共汽车）。只有当必须连接多个列车以提供极大的运输量时，才有铁路运输工具继续存在的必要：如地下铁路或郊区列车。电车不应该再出现在现代城市的市中心。

间隔400米的地块大约有16公顷用地，根据商业区或住宅区的不同，人口分为50000人或6000人。很自然地，可以保持巴黎地铁的平均站距，在所划分的每块地块的中心位置各设一个地铁站。

在城市的两条轴线上，快速干道的下面一层为穿过式地铁，连通花园新城所在市郊的4个端点，从而构成地下铁路网的汇集管道（见下一章）。两条庞大的快速干道的下面还包含有：地下二层的单向交通（环形）的郊区直达火车；地下三层的服务外省的四条主要铁路线，尽端式的铁路线，或是与环形交通系统相连接的铁路线。

车站

只有一座车站。车站只能位于市中心。这是它唯一的所在地；没有任何理由将其分配至他处，车站就像轮轴一样。

车站应该特别地建于地下。其在城市地面以上的两层楼的屋顶可建造为出租飞机使用的机场。出租飞机机场（与保护区内的主要机场相联系）必须与地铁、郊区铁路、外省铁路、"快速干道"以及交通运输管理部门等密切联系（见下一章的"车站规划"）。

……

花园新城

美学、经济学、完美性、现代精神

一句话就能够概括未来的需要：必须在开敞的空间内建设。规划方案必须以一种纯粹几何学的方式进行，包含许多精细的安排。

现状的城市因为其非几何性而濒临垂死的边缘。在开敞的空间内建设是希望通过统一的规划来取代目前的紊乱局面。除此之外没有其他办法。

几何学规划的结果，标准化生产。

标准化生产的结果：规范、完美性（类型的

natural thing, therefore, would be to continue to apply our unit of distance as it exists in the Paris tubes 10-day (namely, 400 yards) and to put a station in the middle of each plot.
Following the two great axes of the city, two "storeys" below the arterial roads for fast traffic, would run the tubes leading to the four furthest points of the garden suburbs, and linking up with the metropolitan network (see the next chapter). At a still lower level, and again following these two main axes, would run the one-way loop systems for suburban traffic, and below these again the four great main lines serving the provinces and running north, south, east and west. These main lines would end at the Central Station, or better still might be connected up by a loop system.

THE STATION

There is only one station. The only place for the station is in the centre of the city. It is the natural place for it, and there is no reason for putting it anywhere else. The railway station is the hub of the wheel.
The station would be an essentially subterranean building. Its roof, which would be two storeys above the natural ground level of the city, would form the aerodrome for aero-taxis. This aerodrome (linked up with the main aerodrome in the protected zone) must be in close contact with the tubes, the suburban lines, the main lines, the main arteries and the administrative services connected with all these. (See the plan of the Station in the following chapter.)
…

GARDEN CITIES

THEIR AESTHETIC, ECONOMY, PERFECTION AND MODERN OUTLOOK.
A simple phrase suffices to express the necessities of to-morrow: WE MUST BUILD IN THE OPEN. The lay-out must be of a purely geometrical kind, with all its many and delicate implications.
The city of to-day is a dying thing because it is not geometrical. To build in the open would be to replace our present haphazard arrangements, which are all we have to-day, by a uniform lay-out. Unless we do this there is no salvation.
The result of a true geometrical lay-out is repetition.
The result of repetition is a standard, the perfect form (i.e. the creation of standard types). A geometrical lay-out means that mathematics play their part. There is no first-rate human production but has geometry at its base. It is of the very essence of Architecture. To introduce uniformity into the building of the city we must industrialize building. Building is the one economic activity which has so far resisted industrialization.
It has thus escaped the march of progress, with the result that the cost of building is still abnormally high.
The architect, from a professional point of view, has become a twisted sort of creature. He has grown to love irregular sites, claiming that they inspire him with original ideas or getting round them. Of course he is wrong. For nowadays the only building that can be

undertaken must be either for the rich, or built at a loss (as, for instance, in the case of municipal housing schemes), or else by jerry-building and so robbing the inhabitant of all amenities. A motor-car which is achieved by mass production is a masterpiece of comfort, precision, balance and good taste. A house built to order (on an "interesting" Site) is a masterpiece of incongruity-a monstrous thing. . .

If the builder's yard were reorganized on the lines of standardization and mass production we might have gangs of workmen as keen and intelligent as mechanics.

The mechanic dates back only twenty years, yet already he forms the highest caste of the working world.

The mason dates…from time immemorial! He bangs away with feet and hammer. He smashes up everything round him, and the plant entrusted to him falls to pieces in a few months. The spirit of the mason must be disciplined by making him part of the severe and exact machinery of the industrialized builder's yard.

The cost of building would fall in the proportion of 10 to 2.

The wages of the labourers would fall into definite categories; to each according to his merits and service rendered.

The "interesting" or erratic site absorbs every creative faculty of the architect and wears him out. What results is equally erratic: lopsided abortions; a specialist's solution which can only please other specialists.

We must build in the open: both within the city and around it.

Then having worked through every necessary technical stage and using absolute ECONOMY, we shall be in a position to experience the intense joys of a creative art which is based on geometry.

THE CITY AND ITS AESTHETIC
(The plan of a city which is here presented is a direct consequence of purely geometric considerations.)

A new unit on a large scale (400 yards) inspires everything. Though the gridiron arrangement of the streets every 400 yards (sometimes only 200) is uniform (with a consequent ease in finding one's way about), no two streets are in any way alike. This is where, in a magnificent contrapuntal symphony, the forces of geometry come into play.

Suppose we are entering the city by way of the Great Park. Our fast car takes the special elevated motor track between the majestic sky-scrapers: as we approach nearer there is seen the repetition against the sky of the twenty-four sky-scrapers; to our left and right on the outskirts of each particular area are the municipal and administrative buildings; and enclosing the space are the museums and university buildings.

Then suddenly we find ourselves at the feet of the first sky-scrapers. But here we have, not the meagre shaft of sunlight which so faintly illumines the dismal streets of New York, but an immensity of space. The whole city is a Park. The terraces stretch out over lawns and into groves. Low buildings of a horizontal kind lead the eye

建立）。几何学的规划意味着将几何学运用于建筑工程之中。优秀的人类作品中没有不应用到几何学的。几何学就是建筑学的精髓。为了将标准化引入城市建设之中，首先必须要工业化的建筑施工。建筑施工至今仍是唯一一项躲避着工业化的经济活动。建筑施工因此而脱离了进步。所以它依然造价昂贵。

建筑师在专业上已误入歧途。他开始喜爱怪异的地形，企图从中寻找获得原创性解决办法的奥秘。建筑师已步入歧途。我们今后就只能为富人而建或者以亏损的方式（如以市政当局的预算）而建，或者是，不顾一切地粗制滥造，不考虑居民必须的舒适性。一辆批量生产的汽车是舒适、精准、均衡及高品位之杰作。一栋定做的房屋（在怪异的地形上）是一件不合时宜的杰作——一个怪物。

如果建筑施工能够实现工业化，我们便可以组成像机械工程师团队那般敏锐机智的工匠团队。

机械工程师起源于 20 年前，在工匠领域中属于最高阶层。

泥瓦匠……历来始终都有吧！他用双脚和榔头敲敲打打，他破坏掉周围的一切，交给他的工具设备在几个月内就消耗殆尽。必须革新泥瓦匠的精神，使其以一种严密而精准的工业化建筑施工方式而工作。

建筑成本也可由 10 降至 2。

建筑施工的人工报酬，根据泰勒理论：应按照每人所提供的服务应得的奖赏而加以区分。

怪异的地形耗尽建筑师的所有创作才能并使他们精疲力竭，以此种方式而产生的作品也是怪异的——根据其特征而言——是个瘸腿的早产儿，特别的解决办法只能取悦那种特别的人。必须在开敞的空间内建设：城里、城外。

以绝对经济的方式完成了各个必须的（技术）阶段，即能感受到那种基于几何学的创造性艺术之强烈喜悦。

城市美学

（此处所设计之城市为纯粹几何学推论之结果。）

新的大模距（400 米）给一切都带来了活力。划分为 400 米或 200 米的道路方格网整齐划一（便于居民辨识方位），但彼此之外貌并不相同。

在这里，在一片华丽的交响乐中，几何学的威力正在奏效。

我们从英式花园而进入城市。汽车沿着高架的城市干道高速行驶：雄伟的摩天大楼之间的通道。我们逐渐接近：24栋摩天大楼的标准化空间；左右两侧的深处是公共服务设施；其周围则是博物馆和大学的建筑。

我们很快地就来到了第一座摩天大楼的跟前。大楼之间不是像纽约那种令人焦虑的狭缝，取而代之的是宽敞的空间，公园向四方延伸，露台层层排列于草坪与绿树成荫之处。低矮的建筑物引导着人们的视线至远处起伏的树林。那些微小的检察官楼哪去了呢？就在这个充满宁静和洁净空气、聚满人群的城市之中，一切的嘈杂喧哗均被埋藏于绿树成荫之下。混乱的纽约被征服了。展现于眼前的，沐浴于日光之中的，正是现代城市。

汽车降低了100公里的时速而下了高架桥；它缓缓地驶入了住宅社区。锯齿的形状将建筑的透视拉向远方，花园、游乐场、运动场，到处可见晴朗的天空，无限开阔。空中花园镶边的绿色，将露台的屋顶水平线剪切成简洁的形状。细节的一致性强调出了一片大体量的实体轮廓线。通过远距离蓝色的柔和化，摩天大楼立起巨大的几何形体玻璃幕墙。被玻璃包裹着的立面上，映射出蔚蓝的天空，真是令人着迷啊。立面硕大无比，但却光彩夺目。

四周的景色各不相同；400米的方格，但它被建筑的处理手法巧妙地改变着！（锯齿则是600米×400米的模距）

搭乘飞机从君士坦丁堡，或许从北京出发，来到这里的旅客，在河流和森林的混乱轮廓之中会突然看到，此处呈现出一片人类光辉城市建设的显著标志：人类智慧的显著标志。

黄昏降临之时，摩天大楼的玻璃中开始闪闪发光。

这并不是充满危险的未来主义，不是猛烈地投向观众的文学炸弹。这是利用建筑学的造型资源来组织光线所创造的景致。

《明日城市》PP156-166
勒·柯布西耶著，李浩译
中国建筑工业出版社，2009年3月第1版

on to the foliage of the trees. Where are now the trivial Procuracies? Here is the CITY with its crowds living in peace and pure air, where noise is smothered under the foliage of green trees. The chaos of New York is overcome. Here, bathed in light, stands the modern city. Our car has left the elevated track and has dropped its speed of sixty miles an hour to run gently through the residential quarters. The "set-backs" permit of vast architectural perspectives. There are gardens, games and sports grounds. And sky everywhere, as far as the eye can see. The square silhouettes of the terraced roofs stand clear against the sky, bordered with the verdure of the hanging gardens. The uniformity of the units that compose the picture throw into relief the firm lines on which the far-flung masses are constructed. Their outlines softened by distance, the sky-scrapers raise immense geometrical facades all of glass, and in them is reflected the blue glory of the sky. An overwhelming sensation. Immense but radiant prisms. And in every direction we have a varying spectacle: our "gridiron" is based on a unit of 400 yards, but it is strangely modified by architectural devices! (The "set-backs" are in counterpoint, on a unit of 600 x 400.)
The traveler in his airplane, arriving from Constantinople or Pekin it may be, suddenly sees appearing through the wavering lines of rivers and patches of forests that clear imprint which marks a city which has grown in accordance with the spirit of man: the mark of the human brain at work.
As twilight falls the glass sky-scrapers seem to flame.
This is no dangerous futurism, a sort of literary dynamite flung violently at the spectator. It is a spectacle organized by an Architecture which uses plastic resources for the modulation of forms seen in light.

The City of To-morrow and Its Planning, PP164-178
Le Corbusier
Translated from the 8th French Edition of Urbanisme by Frederick Etchells
Dover Publications, Inc., New York

功能混合

Although it is hard to believe, while looking at dull gray areas, or at housing projects or at civic centers, the fact is that big cities are natural generators of diversity and prolific incubators of new enterprise and ideas of all kinds. Moreover, big cities are the natural economic homes of immense numbers and ranges of small enterprises.

The principal studies of variety and size among city enterprises happen to be studies of manufacturing, notably those by Raymond Vernon, author of *Anatomy of a Metropolis*, and by P. Sargant Florence, who has examined the effect of cities on manufacturing both here and in England.

Characteristically, the larger a city, the greater the variety of its manufacturing, and also the greater both the number and the proportion of its small manufacturers. The reasons for this, in brief, are that big enterprises have greater self-sufficiency than small ones, are able to maintain within themselves most of the skills and equipment they need, can warehouse for themselves, can sell to a broad market which they can seek out wherever it may be. They need not be in cities, and although sometimes it is advantageous for them to be there, often it is more advantageous not to. But for small manufacturers, everything is reversed. Typically they must draw on many and varied supplies and skills outside themselves, they must serve a narrow market at the point where a market exists, and they must be sensitive to quick changes in this market. Without cities, they would simply not exist. Dependant on a huge diversity of other city enterprises, they can add further to that diversity. This last is a most important point to remember. City diversity itself permits and stimulates more diversity.

For many activities other than manufacturing, the situation is analogous. For example, when Connecticut General Life Insurance Company built a new headquarters in the countryside beyond Hartford, it could do so only by dint of providing - in addition to the usual working spaces and rest rooms, medical suite and the like - a large general store, a beauty parlor, a bowling alley, a cafeteria, a theater and a great variety of games space. These facilities are inherently inefficient,

当我们的眼光扫过那些单调灰色的地带，或者是一些廉租住宅区，或者是一些市民中心等等时，我们很难相信这么一个事实，即大城市是天然的多样化的发动机，是各种各样新思想和新企业的孵化器。但是事实就是如此。是一步说，大城市是成千上万个各个行业的小企业的天然经济家园。

对城市企业的种类和大小的最主要的研究，恰恰就是关于制造业方面的研究，最著名的是雷蒙德·弗农，《解剖大都市》的作者和P．萨金特·弗洛伦斯，他考察美国及英国的城市对制造业的影响。

一个明显的特征是，城市越大，制造业的种类就越多，同时小制造业主的数量和比例也就越大。简单地说，原因是大企业有更多的自给自足的能力，它们能够在企业内部获得它们需要的大部分技术和设备，能够自己贮存这些东西，能够把这些技术和设备远销到它们能够达到的地方。它们并不一定要在城市里，尽管有时候在城市里是一种优势，但往往情况是不在城市里反而更好。但是对小制造业主来说，情况刚好相反。一种典型的情况是，它们必须依赖其本身以外的技术供应，它们必须服务于一个狭小的市场，而且这个市场还必须要存在。它们还必须要对这个市场的变化非常敏感。没有城市，它们就根本不可能存在。它们依赖于城市中的其他形式各样的企业，同时也为这种多样性添砖加瓦。这一点是最应该让我们记住的地方。城市的多样性本身就会带来和刺激更多的多样性的产生。

对制造业以外的经济活动而言，情况也是类似的。比如，康涅狄格州的通用人寿保险公司在哈特福德（州首府）郊区的乡村建一个新的总部，他们除了提供通常的工作场所和休息室、医疗室等诸如此类的东西外，还不得不修建一个大百货店、一家美容院、一家保龄球馆、一个餐厅、一个剧院和一个玩各种游戏的游艺厅。这种设施注定是无效的，大部分时间里空闲着，需要用补贴来支撑它们，这倒不是因为它们天生就该赔本，而是因为它们的用途不能充分发挥出来。但是，另一方面，它们肩负着要维护职员生存的责任而不得不拼命挣扎。大企业可以支撑这些奢华但注定是无效的支出，用其在别的地方获得的盈利来填补亏损。但是小企业就根本无能为力。如果它们要为员工获得一样的或更好的条件，就必须背靠一个活跃的城市，它们的员工在这里可以找到各类他们需要的方便和选择。事实上，有很多说法认为为什么战后将会产生大公司从城市迁往郊区的潮流，但最后仅仅流于空谈而已。为什么？且不说其他很多原因，其中一个是在郊区土地和空间上所获得的差价被支付员工所需设施占据的更大的空间费用抵消了，而这些设施在城市里则根本不需要，也用不着需要专门的雇员或顾客来提供支持。为什么大企业一直呆在城市里，就像和小企业一样，另一个原因是很多职员，特别是经理们需要和企业以外的人进行密切的、面对面的接触和交流——包括来自小企业的人员。

　　城市可以给小单位提供的好处在零售、文化设施和娱乐方面也同样明显。这是因为城市人口众多，足以满足这些单位的各类不同的选择。当然，我们会发现大型的企业（商场）在小型的住宅区域也有优势。比如，城镇和郊区是巨型超市的天然家园，但对杂货店、标准电影院和剧院来说却并不如此。一个简单的原因是没有这么多人来光顾这些不同种类的商业场所，尽管只要这些场所存在也会有人去，但人数很少。但是，相比之下，城市却是超市、标准电影院（带小吃店）、维也纳面包店、异域杂货店、艺术电影院等的天然家园。所有这一切都是共生共存，标准的和奇特的、大型的和小规模的互相依存。城市里那些热闹、活跃的地方，一般规模小的地方比大型场所的地方人多。就像小型的制造业主们一样，缺少了城市这个环境，这些小商家就不可能在任何地方生存。没有城市，就没有它们。

　　城市里的多样性，不管是什么样的，都与一个事实有关，即，城市拥有众多人口，人们的兴趣、品位、需求、感觉和偏好五花八门、千姿百态。

　　即使一些标准的、但小规模的如单人经营的五金店、杂货店、糖果店和酒吧，在一些城市的热闹街区生意也能够做得很好，因为这些地方有足够多的人在短时间内可以非常方便地对它们提供支持，反过来，它们给人们提供的方便也成了这些小商家保持生意兴隆的一个重要的潜在原

idle most of the time. They require subsidy, not because they are kinds of enterprises which are necessarily money losers, but because here their use is so limited. They were presumed necessary, however, to compete for a working force, and to hold it. A large company can absorb the luxury of such inherent inefficiencies and balance them against other advantages it seeks. But small offices can do nothing of the kind. If they want to compete for a work force on even terms or better, they must be in a lively city setting where their employees find the range of subsidiary conveniences and choices that they want and need. Indeed, one reason, among many others, why the much-heralded postwar exodus of big offices from cities turned out to be mostly talk is that the differentials in cost of suburban land and space are typically canceled by the greater amount of space per worker required for facilities that in cities no single employer need provide, nor any one corps of workers or customers support. Another reason why such enterprises have stayed in cities, along with small firms, is that many of their employees, especially executives, need to be in close, face-to-face touch and communication with people outside the firm - including people from small firms.

The benefits that cities offer to smallness are just as marked in retail trade, cultural facilities and entertainment. This is because city populations are large enough to support wide range of varieties and choice in these things. And again we find that bigness has all the advantages in smaller settlements. Towns and suburbs, for instance, are natural homes for huge supermarkets and for little else in the way of groceries, for standard movie houses or drive-ins and for little else in the way of theater. There are simply not enough people to support further variety, although there may be people (too few of them) who would draw upon it were it there. Cities, however, are the natural homes of supermarkets and standard movie houses plus delicatessens, Viennese bakeries, foreign groceries, art movies, and so on, all of which can be found co-existing, the standard with the strange, the large with the small. Wherever lively and popular parts of cities are found, the small much outnumber the large. Like the small manufacturers, these small enterprises would not exist somewhere else, in the absence of cities. Without cities, they would not exist.

The diversity, of whatever kind, that is generated by cities rests on the fact that in cities so many people are so close together, and among them contain so many different tastes, skills, needs, supplies, and bees in their bonnets.

Even quite standard, but small, operations like proprietor-and-one-clerk hardware stores, drug stores, candy stores and bars can and do flourish in extraordinary numbers and incidence in lively districts of cities because there are enough people to support their presence at short, convenient intervals, and in turn this convenience and neighborhood personal quality are big parts of such enterprises' stock in trade. Once they are unable to be supported at close convenient intervals, they lose this advantage. In a given geographical territory, half as many people will not support half as many such enterprises spaced at twice the distance. When distance inconvenience sets in, the small, the various and the personal wither away.

As we have transformed from a rural and small-town country into an urban country, business enterprises have thus become more numerous, not only in absolute terms, but also in proportionate terms. In 1900 there were 21 independent nonfarm businesses for each 1,000 persons in the total U.S. Population. In 1959, in spite of the immense growth of giant enterprises during the interval, there were 26% independent nonfarm businesses for each 1,000 persons in the population. With urbanization, the big get bigger, but the small also get more numerous.

Smallness and diversity, to be sure, are not synonyms. The diversity of city enterprises includes all degrees of size, but great variety does mean a high proportion of small elements. A lively city scene is lively largely by virtue of its enormous collection of small elements.

Nor is the diversity that is important for city districts by any means confined to profit-making enterprises and to retail commerce, and for this reason it may seem that I put an undue emphasis on retail trade. I think not, however. Commercial diversity is in itself, immensely important for cities, socially as well as economically. Most of the uses of diversity on which I dwelt in Part I of this book depend directly or indirectly upon the presence of plentiful, convenient, diverse city commerce. But more than this, wherever we find a city district with an exuberant variety and plenty in its commerce, we are apt to find that it contains a good many other kinds of diversity also, including variety of cultural opportunities, variety of scenes, and a great variety in its population and other users. This is more than coincidence. The same physical and economic conditions that generate diverse commerce are intimately related to the production, or the presence, of other kinds of city variety.

But although cities may fairly be called natural economic generators of diversity and natural economic incubators of new enterprises, this does not mean that cities automatically generate diversity just by existing. They generate it because of the various efficient economic pools of use that they form. Wherever they fail to form such pools of use, they are little better, if any, at generating diversity than small settlements. And the fact that they need diversity socially, unlike small settlements makes no difference. For our purposes here, the most striking fact to note is the extraordinary unevenness with which cities generate diversity.

On the one hand, for example, people who live and work in Boston's North End, or New York's Upper East Side or San Francisco's North Beach-Telegraph Hill, are able to use and enjoy very considerable amounts of diversity and vitality. Their visitors help immensely. But the visitors did not create the foundations of diversity in areas like these, nor in the many pockets of diversity and economic efficiency scattered here and there,

因。一旦它们不能得到人们这种短时间内的方便的支持，它们就失去优势。在任何一个城市地区，人数减少一半，相应来说，即使店家的数量也减少一半，但如果互相间距离增加一倍，它们就很难生存。一旦产生了距离上的不方便，这些小型的、形式各样的、人情味很足的场所 就会萎靡不振。

随着我们从乡村和小城镇的国家转向一个城市化的国家，商业和企业不仅在绝对数量上大幅增加，而且在比例上也有很大的增加。1900年，在整个美国的人口中，每1000人只有21家独立的非农场企业。在1959年，尽管大企业增长迅速，但在每1000人中仍有26.5家独立的非农场企业。随着都市化进程的加快，大企业越来越大，但小企业在数量上也越来越多。

当然，要说明的是，小型和多样不是同义词。城市企业的多样性包括各种大小程度不同的企业。但是，种类的繁多在很大程度上确实是因为有小的因素存在。一个有着热闹活跃景致的城市，其中一个很重要的原因是有许许多多的小型东西的存在。

另外，城市街区的多样性也不仅是局限于有利润的企业和零售业。从这个方面讲，似乎我对零售业强调过多，其实不是。对城市来说，商业上的多样性无论是从经济上还是从社会的角度对城市都有极其重要的影响。我在本书第一章提到的多样性的各种用途都与众多的、方便的和多样化的城市商业有直接或间接的关系。但是，实际情况不止于此，每当我们发现一个城市地区的商业生活丰富多彩，我们就同时会发现这个地方也拥有很多其他的丰富多彩的生活，如各种各样的文化机会、多彩的街头景致等。此外，人口和其他的使用者也呈现出多样化的色彩。这并不是巧合。促成商业多样性的物质和经济条件与城市的其他的多样性的存在和产生紧密相关。

但是，尽管城市被非常恰当地称为是经济多样性的天然发动机和新企业的天然孵化器，这并不表明城市只要通过它的存在就能自动生发多样性。城市之所以能够生发多样性，是因为它们能够集中各种有效的经济资源。一旦做不到这一点，在生发多样性方面它们比一些小城镇好不了多少。尽管它们在社会方面也需要多样性——这与小城镇不同，但这个事实也并不能说明多少问题。就我们现在讨论的目的而言，一个最突出的问题是城市在生发多样性方面的不平衡。

比如，一方面，在波士顿北端或纽约的上东区或旧金山的北滩——电报山，人们能够使用或拥有足够多的多样性和地区的活力。这里的来访者对此提供了很大的帮助。但是，这些地区的多样性的基础并不是由来访者创立的，同样，很多分布得很零散的多样性和经济效益茂盛的地方也不光是由来访者促成的。这些地方的发展有时候出人意料，在大城市尤其如此。哪儿有活力，来访者就往哪儿走，他们在分享这里的多样性，同时也为这里的多样性加了一把力。

另一方面，一些居住地区人口众多，但并不能产生任何有用的效应，有的只是一派死气沉沉，以致最后大家对那个地方只有强烈的厌恶感。这并不是因为这些地方的居民与众不同，比别的地方的要迟钝一些，或对活力和多样性没有感觉，或他们是一些嗅觉特别敏感的人，常常跑到其他充满活力的地方去，而是因为这些地区出了问题，缺少能够把这个地区的人口优势催化成能跟经济互动的因素，并且以此来组成有效的使用资源。

显然，在一个城市地区，能够成为城市人口一员的人数没有一个限度，因为这样的人口数量很可能会被浪费掉，发挥不了作用。举个例子来说，纽约的布朗克斯区拥有大约1,500,000的人口，但令人不解的是，布朗克斯区却不见城市的活力、多样性和吸引力。应该说明的是，此地的居民都还是很忠诚于这个地区的，主要是依恋于这个"古老街区"零零散散地表现出活力的街道生活，可惜这样的情况太少了。

一个拥有1,500,000人口的地区竟然连开办一些能吸引人的饭店这样的具有多样性因素的简单事情都做不到。《纽约名胜和玩乐》这本导游手册的作者凯特·西蒙，介绍了几百家饭店和其他商业场所，特别是开在城市某些意想不到的地方的这些场所。她这样做并不是出于势利，而是真正想给读者介绍一些她发现的便宜的地方。但是尽管西蒙小姐费了很大的劲，她还是不得不放弃把布朗克斯这个大号地区作为一个值得介绍的地方推荐给读者，因为在价格上没有优势可言。在介绍了布朗克斯区的动物园和植物园这两个不得不提的名胜点，并对其表示了一番尊重后，她百般无奈地推荐了动物园外面惟一一家可吃饭的地方。就这么一个地方，她还是向读者表示了很大的歉意："可惜的是，这儿的街区逐渐缩小，最后伸向一个无人区，这个餐馆也应该再稍稍装饰一下，但是让人感到欣慰的是……布朗克斯区最好的医疗人员很可能就围坐在你的身边。"

这就是布朗克斯，这个地区变成这样真是太不幸了；对于现在住在那儿的人来说真是太不幸了，对于那些因为经济原因别无他选而在未来不

sometimes most unexpectedly, in big cities. The visitors sniff out where something vigorous exists already, and come to share it, thereby further supporting it.

At the other extreme, huge city settlements of people exist without their presence generating anything much except stagnation and, ultimately, a fatal discontent with the place. It is not that they are a different kind of people, somehow duller or unappreciative of vigor and diversity. Often they include hordes of searchers, trying to sniff out these attributes somewhere, anywhere. Rather, something is wrong with their districts; something is lacking to catalyze a district population's ability to interact economically and help form effective pools of use.

Apparently there is no limit to the numbers of people in a city whose potentiality as city populations can thus be wasted. Consider, for instance, the Bronx, a borough of New York containing some one and a half million people. The Bronx is woefully short of urban vitality, diversity and magnetism. It has its loyal residents, to be sure, mostly attached to little bloomings of street life here and there in "the old neighborhood," but not nearly enough of them.

In so simple a matter of city amenity and diversity as interesting restaurants, the 1,500,000 people in the Bronx cannot produce. Kate Simon, the author of a guidebook, *New York Places and Pleasures*, describes hundreds of restaurants and other commercial establishments, particularly in unexpected and out-of-the-way parts of the city. She is not snobbish, and dearly likes to present her readers with inexpensive discoveries. But although Miss Simon tries hard, she has to give up the great settlement of the Bronx as thin pickings at any price. After paying homage to the two solid metropolitan attractions in the borough, the zoo and the Botanical Gardens, she is hard put to recommend a single place to eat outside the zoo grounds. The one possibility she is able to offer, she accompanies with this apology: "The neighborhood trails off sadly into a no man's land, and the restaurant can stand a little refurbishing, but there's the comfort of knowing that…the best of Bronx medical skill is likely to be sitting all around you."

Well, that is the Bronx, and it is too bad it is so; too bad for the people who live there now, too bad for the people who are going to inherit it in future out of their lack of economic choice, and too bad for the city as a whole.

And if the Bronx is a sorry waste of city potentialities, as it is, consider the even more deplorable fact that it is possible for whole cities to exist, whole metropolitan areas, with pitifully little city diversity and choice. Virtually all of urban Detroit is as weak on vitality and diversity as the Bronx. It is ring superimposed upon ring of failed gray belts. Even Detroit's downtown itself cannot produce a respectable amount of diversity. It is dispirited and dull, and almost deserted by seven o'clock of an evening.

So long as we are content to believe that city diversity represents accident and chaos, of course its erratic generation appears to represent a mystery.

However, the conditions that generate city diversity are quite easy to discover by observing places in which diversity flourishes and studying the economic reasons why it can flourish in these places. Although the results are intricate, and the ingredients producing them may vary enormously, this complexity is based on tangible economic relationships which, in principle, are much simpler than the intricate urban mixtures they make possible.

To generate exuberant diversity in a city's streets and districts, four conditions are indispensable:

1. The district, and indeed as many of its internal parts as possible, must serve more than one primary function; preferably more than two. These must insure the presence of people who go outdoors on different schedules and are in the place for different purposes, but who are able to use many facilities in common.
2. Most blocks must be short; that is, streets and opportunities to turn corners must be frequent.
3. The district must mingle buildings that vary in age and condition, including a good proportion of old ones so that they vary in the economic yield they must produce. This mingling must be fairly close-grained.
4. There must be a sufficiently dense concentration of people, for whatever purposes they may be there. This includes dense concentration in the case of people who are there because of residence.

The necessity for these four conditions is the most important point this book has to make. In combination, these conditions create effective economic pools of use. Given these four conditions, not all city districts will produce a diversity equivalent to one another. The potentials of different districts differ for many reasons; but, given the development of these four conditions (or the best approximation to their full development that can be managed in real life), a city district should be able to realize its best potential, wherever that may lie. Obstacles to doing so will have been removed. The range may not stretch to African sculpture or schools of drama or Rumanian tea houses, but such as the possibilities are, whether for grocery stores, pottery schools, movies, candy stores, florists, art shows, immigrants' clubs, hardware stores, eating places, or whatever, they will get their best chance. And along with them, city life will get its best chances.

The Death and Life of Great American Cities, PP145-151
Jane Jacobs
Vintage Books Edition, December 1992
Random House, Inc., New York

得不接受这个地方的人来说真是太不幸了，最后对整个城市来说也真是太不幸了。

布朗克斯区浪费了城市的潜在资源，如果说这是一个遗憾的话，那么一个更加悲惨的事实是，我们所有的城市，全部的大都市区域都有可能落到这种地步——多样性和选择少得可怜。整个底特律城市区域在活力和多样性方面几乎和布朗克斯区一样差劲。一个环路接一个环路都是一派衰败的灰色地带。即使是在底特律市中心也找不到多少像样的多样性，只有单调、沉闷，到晚上七点时整个地区空无一人。

如果我们非常乐意相信城市的多样性是随意性很强的、没有什么规律可言的，那么制造这种理论的人其实也在制造一种神话，一种关于多样性无处可寻的神话。

但是，产生城市多样性的条件是很容易发现的。只要观察一下那些多样性蓬勃发展的地方，研究一下产生如此强的多样性的经济原因，就能有所发现。尽管得出的结果是复杂的，产生的原因会有巨大的不同，但是这种复杂性是建立在具体的经济联系上的，而这种经济联系比之因其而产生的城市复杂的综合关系在本质上要简单得多。

要想在城市的街道和地区生发丰富的多样性，四个条件不可缺少：

1) 地区以及其尽可能多的内部区域的主要功能必须要多于一个，最好是多于两个。这些功能必须要确保人流的存在，不管是按照不同的日程出门的人，还是因不同的目的来到此地的人，他们都应该能够使用很多共同的设施。

2) 大多数的街段必须要短，也就是说，在街道上能够很容易拐弯。

3) 一个地区的建筑物应该各色各样，年代和状况各不相同，应包括适当比例的老建筑，因此在经济效用方面可各不相同。这种各色不同建筑的混合必须相当均匀。

4) 人流的密度必须要达到足够高的程度，不管这些人是为什么目的来到这里的。这也包括本地居民的人流也要达到相等的密度。

这四个条件的必要性是本书一个最重要的观点。这四个条件的结合能产生最有效的经济资源。虽然即使有了这四个条件，也不是所有的城市地区都能生发相同的多样性。不同地区的潜能因种种原因而表现不同，但是，只要能在这四个条件方面有所发展（或在实际生活中能做到靠近这个方向发展），那么一个城市的地区不管其位置在于何方，是应该能够发挥其最大潜能的。阻碍这个目标得以实现的障碍将会被消除。也许像非洲雕塑、戏剧学校或罗马尼亚茶馆这样的东西并不是特别需要，但诸如杂货店、陶瓷学校、电影院、糖果店、艺术花店、表演场所、移民俱乐部、五金店、饭店，等等，这些场所都会拥有最佳的发展机会。当然与这些场所一起的还有城市生活，也能得到最好的发展。

《美国大城市的死与生》PP159-166
简·雅各布斯著，金衡山译
译林出版社，2005年5月

类型问题

Since housing, like many other urban issues, concerns cities and, for better or worse, cities are something we can describe, it is useful to approach this issue in the context of a specific city. In speaking of housing in a specific city, then, it is necessary to try to make as few generalizations as possible. Clearly all cities will always have something in common relative to this issue, and by inquiring how much one artifact has in common with others, we will come closer to elaborating a general theory.

The problem of housing typology in Berlin is extremely interesting, especially with respect to other cities, and I will attempt to indicate the patterns that enable us to recognize a certain uniformity or continuity in this issue in Berlin, ultimately showing the capacity of a few typical residential models, past and present, to shed light on a series of questions concerning housing which in turn relate to the urban condition and a theory of urban development. The particular interest of Berlin housing becomes apparent on an examination of the city's plan. In 1936, the geographer Louis Herbert distinguished four major types of structures in Berlin; these distinctions related to four zones defined by their distance from the historical center:

1. a zone of uniform and continuous structures, such as buildings of the "large city" type, possessing at least four stories;
2. a zone of diversified urban structures, which could be divided into two classes: in the center of the city, new buildings mixed with very old and low buildings of no more than three floors; and along the edges of the center, a continuous interspersion of high and low housing, open spaces, fields, and parceled land;
3. large areas for industry; ,
4. residential areas open at the outer edges of the city, comprised of villas and single-family dwellings principally constructed after 1918.

Between the fourth zone and the periphery there was a continuous blending of industrial zones, residential zones, and villages in transformation. These external zones differed greatly from one another, and ranged from the working-class and industrial districts of Henningsdorf and Pankow to the upper-class district

既然同其他许多城市问题一样，住房与城市有关，并且我们总归可以描述城市，因此在具体的城市环境中研究住房问题是有益的。不过，在谈论具体城市中的住房问题时，我们应当尽可能地避免一概而论。所有城市显然在这方面都有一些共同之处，通过研究某一建筑体和其他建筑体的相同之处，我们可以进一步地阐述一种普遍的理论。

柏林的住房类型问题极为有趣，与其他的城市相比，尤其是这样。我将努力揭示那些使我们得以认识这类问题中某种统一性或连续性的形式，表明过去和现在少数典型居住模式的容量，从而澄清一系列与城市条件和城市发展理论相关的住房问题。通过柏林城的规划，我们会更为清楚地认识到其住房的特别意义。1936年，地理学家赫伯特在柏林城分出四种主要结构类型，与距历史中心不等的四个地区相对应：

1．具有统一和连续结构的地区，例如至少有四层高的"大城市"类型的建筑物；

2．具有不同城市结构的地区，它可以分为两类：一类位于市中心，为新建筑物与三层以下的很古老且低矮的建筑物的混合体；另一类位于市中心的边缘，连续散置着高层和低层住宅，开敞空间，田野和小块土地；

3．大片的工业区；

4．城市外缘的住宅区，主要由1918年以后建成的别墅和独户住宅组成。

在第四地区和周边地带之间，有一连续的工业，居住和转变中的村落混合区。这些外围地区彼此之间差别很大，有工人居住区和工业区，也有上流社会区。在这种结构基础上，鲍迈斯特于1870年提出了分区制概念，此概念后被编入普鲁士建筑规范之中。

在柏林地区中，居住建筑群体的形态是很不

相同的，这些没有直接关系的不同群体的特征正是产生于不同的住宅类型：多层住宅，风险投资住宅和独户住宅。这种类型上的多样化表现出相当现代的城市结构，它后来也出现在欧洲的其他城市中，但还没有像柏林城的那么明确。从城市结构和类型结构的两重性来看，居住区类型的多样化是德国大城市的一个主要特征。居住区只能被认为是这些条件作用的结果。

居住建筑群体的结构可根据以下的基本类型进行划分：

1. 居住街区；
2. 双联式住宅；
3. 独户住宅。

由于历史文化和地理上的原因，柏林城出现这些不同类型的频率要大于其他任何一个欧洲城市。在其他德国城市中，长期保存下来的哥特式建筑构成了城市的主要形象，它们直到二次大战才被毁坏。而柏林的哥特式建筑则在 19 世纪末期就完全消失了。

街坊结构产生于 1851 年的治安条例，是最完整的城市土地开发形式之一；在此结构中，住房围绕一系列院落呈内向布置。这类建筑物也构成了汉堡和维也纳这些城市的特征。这种以"出租兵营"而闻名的住房，在柏林大量出现，使城市具有"兵营城市"的特征。

院落式住房是中欧的一种典型住宅，许多现代建筑师在维也纳和柏林都采用了这种形式。内院被改为大型花园，其中包括托儿所和小贩的售货亭。德国理性主义时期的某些最好住宅与这种形式有关。

理性主义者设计的居住区是以独立的结构为特征的，它们反映了某种很值得讨论的科学观点；在完全自由划分土地的基础上，住房的布局取决于朝向，而不是地区的普遍形式。这些独立的住房与街道没有任何联系，而正是这种情况完全改变了 19 世纪城市发展的类型。在这些居住区中，公共绿地是特别重要的。

研究居住单元这个细胞成为设计居住区的关键。所有设计这些居住区并且研究经济住房类型的建筑师，都试图找到最低居住标准的确切形式，即用组织和经济的观点来确定住房单元的最佳尺寸。这是理性主义者关于住房问题研究的最重要的方面之一。

最低住房标准的公式是以某种生活方式（属于假设尽管可以从统计上证实）和某种住所类型之间的静止关系为前提的。这是造成居住区很快

of Grünewald. On the basis of this already existing organization of Berlin, Reinhard Baumeister in 1870 formulated the concept of zoning which later was incorporated in the Prussian building code.

In Greater Berlin the morphology of the residential complexes was thus quite varied; the different complexes not directly linked to one another were characterized by precise building types: multi-story housing, speculative housing, and single-family housing. This typological variety represents a very modern type of urban structure also produced subsequently in other European cities, even though it never achieved such definitive articulation as in Berlin. Considered in its dual aspect of urban structure and typological structure, it is one of the principal characteristics of the German metropolis. The *Siedlungen* are a product of these conditions and must be so judged.

The structure of the residential complexes can be classified according to the following fundamental types:
1. residential blocks;
2. semi-detached houses;
3. single-family houses.

These different types present themselves in Berlin with greater frequency than in any other European city for historical-cultural and geographical reasons. The Gothic building, preserved for a long time in other German cities where it constituted the primary image up until the devastations of the last war, in Berlin had disappeared almost completely by the end of the nineteenth century.

Block structures, derived from the police regulations of 1851, constitute one of the most integral forms of exploitation of the urban land; these were normally designed around a series of courtyards facing the interior facades of the blocks. Buildings of this type were also characteristic of such cities as Hamburg and Vienna. The very large presence in Berlin of this type of housing, known as *Mietkasernan* or "rental barracks," led to its characterization as a "barracks city."

Courtyard housing represents one typical solution in central Europe, and as such was adopted by many modern architects, in Vienna as in Berlin. The courtyards were transformed into large gardens, which came to include nursery schools and vendors' kiosks. Some of the best examples of housing in the German Rationalist period are associated with this form.

The *Siedlungen* of the Rationalists are characterized by *detached structures*, and these represent a highly polemical and scientific position; their layout, which demands a totally free division of the land, depends on solar orientation rather than on the general form of the district. The structure of these detached buildings is completely disengaged from the street, and precisely for this reason totally alters the nineteenth-century type of urban development. In these examples, public green spaces are particularly important.

The study of the cell, of the individual habitable unit, is fundamental with respect to the Siedlungen. All the architects who worked on shaping these residential districts and engaged in the formulation of economical building types sought to find the exact form of

Existenzminimum, the optimum dimensional unit from the point of view of organization and economy. This is one of the most important aspects of the work of the Rationalists on the problem of housing.

We can only suggest that the formulation of *Existenzminimum* presupposed a static relationship between a certain style of life - hypothetical even if statistically verifiable - and a certain type of lodging, and this resulted in the rapid obsolescence of the *Siedlung*. It revealed itself to be a spatial conception that was too particular, too tied to specific solutions to function as a general element available for wide use in housing. *Existenzminimum* is only one aspect of a far more complex problem in which many variables participate.

There is a strong tradition of the *single-family house* in Berlin residential typology. Although this is one of the most interesting aspects of Rationalist residential typology, I will only mention it briefly since it demands a type of study that is parallel to but outside the bounds of our present task. In this context, Schinkel's projects for the Babelsberg castle for Wilhelm I and the castle and Römische Bäder of Charlottenhof take on particular importance. The plan of the Babelsberg castle presents an ordered structure, almost rigid in the organization of its rooms, while its external form is an attempt to relate to the surrounding context, especially the landscape. In this project one can see how the concept of the villa was borrowed and used as a typological model suitable for a city like Berlin. In this sense, Schinkel's work, constituting the transition from neoclassical models to romantic ones, mainly by way of the English country house, offers the basis for the early twentieth-century type of bourgeois villa.

With the spread of the villa as an urban element in the nineteenth century and the disappearance of Gothic and seventeenth-century houses, with the Substitution of ministries at the center and *Miekasernen* in the peripheral zones, the urban morphology of Berlin was profoundly modified. The changing image of Unter den Linden over the centuries is a typical case. The seventeenth-century street is truly a "promenade" under the lime trees: although of different heights, characteristic of central Europe, constructed on narrow and deep lots and revealing formal elements from Gothic building. Houses of this type were characteristic of Vienna, Prague, Zurich, and many other cities; their origins, often mercantile, were linked to the earliest form of the modern city. With the transformation of the cities in the second half of the nineteenth century, these houses disappeared fairly rapidly, either because of building renewal or because of alterations in the use of areas. With their replacement came a profound modification of the urban landscape, often a rigid monumentalization, as in the case of Unter den Linden. For the older type of house was substituted rental housing and the villa.

To Schumacher, the separation between villas and rental barrack zones in the second half of the nineteenth century represented the crisis of urban unity in the central European city. The villa was sited to provide

过时的原因。这种居住区自身表现为一种过于特别且过于与特定设计相联系的空间概念，因而不能广泛地被用于住房设计中。最低居住标准只是由许多因素构成的相当复杂的问题的一个方面。

在柏林的居住建筑类型中，独户住宅有着悠久的传统。虽然这是理性主义住房类型的最为有趣的方面之一，但我只想在此稍稍提及，因为这需要进行一种研究，它既平行但又超出我们现在所探讨的范围。在这方面，申克尔为威廉一世设计的巴伯尔斯贝格府第和所做的夏洛腾霍夫府第与罗马浴场方案具有特殊的意义。巴伯尔斯贝格府第的布局结构严谨有序，房间的组织近乎刻板，但其外部形式却试图与周围环境尤其是自然环境取得联系。从此方案中，人们可以看到，别墅的概念是怎样被借用并发展为适合柏林这类城市的住宅类型原形的。从这个意义上来看，主要以英国乡间住宅为模本的申克尔作品，标志了从新古典模式向浪漫主义模式的转变，成为20世纪初期资产阶级别墅类型的基础。

随着哥特时期和17世纪时期住房的消失，随着别墅作为城市元素在19世纪中的发展，城市中心为政府部门所取代，城市边缘为出租房屋所取代，柏林的城市形态发生了深刻的变化。菩提树大街在几百年中发生的变化，就是一个典型例子。这条17世纪的街道，确实是酸橙树下的"散步大道"：虽然路边的住宅围墙有高有低，但却有一种整体上的建筑统一性。这些建在狭长地块上的资产阶级住房，带有中欧建筑的特征，在形式上与哥特建筑有某种联系。这种类型的住房在维也纳、布拉格、苏黎士和其他许多城市也很有特征；这些常常起源于商业需要的住房与现代城市的最初形式有关。在19世纪下半叶，这些住房随着城市的转变而迅速消失了，这或是由于建筑物的翻新，或是因为地区用途的改变。这类住房的更替，使城市环境发生了通常带有刻板纪念味的重大变化，例如菩提树大街的变化。因为出租房和别墅取代了老式类型的住房。

在舒马赫看来，19世纪下半叶别墅区和出租兵营式住区之间的分离现象，反映了中欧城市中城市统一的危机。别墅与自然更接近，进一步表现了社会象征和社会阶层。别墅不会也不可能成为连续城市形象中的一部分。另一方面，出租住宅因其成为风险投资住房而贬值，再也没能恢复其居住建筑的价值。

然而，即使舒马赫的观点是正确的，我们也应当承认，别墅在导致现代住宅的类型转变中，发挥了主要的作用。柏林的出租兵营式住区与英国的独户住宅没有什么关系，它是一种特殊且在不断发展的城市居住建筑类型。别墅起初是宫殿的简化（如申克尔设计的巴伯尔斯贝格府第），其内部安排和流线组织日趋精细与合理。穆特修斯的研究对柏林来说是重要的；他强调了功能和自由的内部空间，从而根据理性方法发展了英国乡村住宅建筑的设计原则。

值得注意的是，这些类型上的创新，并没能引发建筑上的敏感变化；为适应资产阶级生活方式，建筑内部设计变得更为自由，但伴随这种自由而来的只是更加富于纪念性的建筑形象，和对申克尔模式的僵化，其中的居住建筑和公共建筑之间的差别令人注目。在这方面，穆特修斯于 1900 年前后设计的建筑物很有说服力，他是当时最典型的柏林城的设计者之一。他对现代住宅的偏见也反映在其论著中，这种偏见关注与形式无关的住宅类型结构。他从德国新古典主义中吸取了某些形式，再加上地方传统的典型元素。这与申克尔模式直接对立，在申氏模式中，住较少地依赖具象的元素，而古典类型的设计则与建筑不冲突。

但是，在 19 世纪后期，居住建筑中引入具象元素的做法在当时所有的建筑中是有代表性的。这也许是为了适应已经变化了的社会条件，满足使住宅具有象征意义的愿望。这当然与舒马赫所谈论的城市统一的危机相一致，也出于区别日益增多的各种社会阶级的需要。现代建筑运动中最为著名的建筑师格罗皮乌斯、门德尔松、黑林等在柏林设计的别墅，以一种相当正统的方式发展了这些类型的模式。尽管这些别墅形象发生了重大变化，但它们显然与之前的折中住房模式毫无决裂之意。社会学家应当确立这种转变具象或标记性元素的方法，不过这是同一现象的不同方面的问题。这些现代住宅将折中式别墅的前提发展为最终结果；人们可以从这一点上认识到，为什么像穆特修斯和凡·德·费尔德被尊为大师：正是因为他们创立了一种普遍的模式，尽管这种模式只是转化了英国和佛兰芒的住宅形式。

独户住宅的所有这些方面都体现在居住区中，而居住区因其自身的复合特征似乎最适宜容纳它们，最适宜给某些倾向以新的定义。为了不在理性主义建筑师们所解释的住房问题上纠缠太久，我将列举一些关于 20 世纪 20 年代在柏林建成的实例。这些实例具有原型上的意义，法兰克福和斯图加特也有同样著名的实例。

显然，理性主义的城市理论浓缩于居住区这个概念之中，至少其居住方面是这样。此概念也

a closer relationship to nature, to further both social representation and social division. It refused to be, or was incapable of being, inserted into a continuous urban image. Rental housing, on the other hand, in becoming speculative housing, was degraded and never recovered the value of civic architecture.

Nonetheless, even if Schumacher's vision is correct, it must be acknowledged that the villa played a large role in the typological transformations that led to the modern house. The Berlin rental barracks have little to do with the single-family English house, whose definition is that of a particular urban type and a continuously developing residential type. The villa was initially a reduction of the palazzo (as in the case of Schinkel's Babelsberg castle), and it became increasingly elaborated in its internal organization and the rationalization and distribution of its circulation. The work of Hermann Muthesius is important for Berlin; by focusing on the function and the freedom of internal spaces, he developed the principles of the English country house in a rational way and in the context of building.

It is significant that these typological innovations did not also lead to sensitive architectural modifications, and that the greater internal freedom - a response to the bourgeois way of life - was only accompanied by a more monumental image of building and an ossification of the Schinkelesque models, wherein the difference between residential architecture and public buildings became marked. In this sense the buildings of Muthesius, one of the most typical builders of urban Berlin around 1900, are illustrative. His preoccupations with the modern house, as also expressed in his theoretical writings, concerned its typological structure independent of its formal aspects. For the latter he accepted a sort of Germanic neoclassicism with the addition of typical elements from the local traditions. This was in direct contrast to Schinkel's models, in which the house was less dependent on representational elements and classical typological schemes were not in conflict with the architecture.

But the introduction of representational elements into residential architecture in the late nineteenth century is typical of all the architecture of the period; it probably corresponds to changed social conditions and the desire to endow the house with an emblematic significance. Certainly it corresponds to the crisis of urban unity of which Schumacher spoke, and thus to the need for differentiation within a structure where increasingly diverse and antagonistic social classes lived. The villas built by the most famous architects of the Modern Movement in Berlin - Gropius, Erich Mendelsohn, Hugo Häring, etc. - developed these typological models in a fairly orthodox way; there was clearly no sense of rupture with their previous eclectic housing models, even if the image of these villas was transformed profoundly. Sociologists will have to establish the way in which this representational or emblematic element was transformed, but it is obviously a question of different aspects of the same phenomenon. These modern houses carry the premises of the eclectic villa to its ultimate consequences, and from his standpoint one can understand why architects like Muthesius and Van de Velde were looked upon as masters: precisely because they established a general model, even if only by translating English or Flemish experiences.

All these themes of the single-family house are represented in the *Siedlung*, which by virtue of its composite character seems to have been suited best to accepting them and to giving certain tendencies a new definition. Without lingering too long on the housing problem as interpreted by the Rationalist architects, I would like to illustrate some examples realized in Berlin during the 1920s. These are prototypical, although one could look to the equally famous examples in Frankfurt and Stuttgart.

Clearly Rationalist urban theory is epitomized, at least with respect to the residential aspect of the problem, in the *Siedlung*, which is probably a sociological model even before it is a spatial one; certainly when we speak of Rationalist urbanism we are thinking of the urbanism of the residential district. This attitude, however, particularly in view of its methodological implications, immediately reveals its insufficiency. To see the urbanism of Rationalism only as the urbanism of the residential district means to limit the magnitude of this experience to German urbanism of the 1920s. In fact, there are so many and such varied solutions that the definition is not even valid for the history of German urbanism. Moreover, the term *residential district*, which as a translation of the German *Siedlung* is as imprecise as it is useful, means so many different things that it is preferable not to use it until we have first examined it carefully.

It is therefore necessary to study actual conditions and artifacts; and given the morphology of Berlin, its richness and the particularity of its urban landscape, the importance of its villas, and so on, it is possible to conclude that here the *Siedlung* has its own special coherence. The close similarity between such *Siedlungen* as Tempelhofer Felde and Britz, or anywhere that the transformation from the English model is evident, renders our primary reference to the urban site more apparent. While such examples as the Friedrich Ebert are closely linked to Rationalist theoretical formulations, in all cases it is difficult to go back from these actual images to an ideology of the *Siedlung*.

Thus, while we have so far considered the *Siedlung* in itself without referring to, indeed ignoring, the context in which it was produced, an analysis of the urbanism of the *Siedlung*, which essentially means the housing problem in Berlin during the 1920s, can only be undertaken with reference to the 1920 plan of Greater Berlin. What was the basis of this plan? It is far closer to certain recent models than one might imagine. In general, the choice of housing was more or less independent of location; it manifests itself as a moment in an urban system which depended on the evolution of a transportation system that in itself embodied the pulse of the city. Through zoning, it encouraged the self-formation of the center as a governmental and

许先是一种社会学的模式，然后才是一种空间的模式。谈到理性主义的城市化，我们便自然地会想到居住区的城市化。但这种态度立刻表现出自身的不足，从其方法论意义的角度来看，更是如此。如果把理性主义的城市化仅仅视为居住区城市化，我们就会把城市化的经验局限于20世纪20年代时的德国城市化之中。事实上，很多不同的城市化经历使得理性主义的城市化定义甚至与德国的城市化历史不相符合。从德文"Siedlung"一字翻译而来的居住区这词虽然有用但并不准确，它所包含的不同内容是如此之多，以至于我们最好在对它进行仔细考察之后再使用它。

所以，我们有必要去研究实际情况和建筑体。从柏林城的形态、城市环境的丰富和特殊性以及别墅的重要性，我们可以得出如下的结论："Siedlung"在此有着特别的含义。腾玻尔霍夫·费尔德和布里茨这类居住区之间的十分相像和那些明显带有英国模式印迹的居住区，明确地表现出对城市地点参照的重要性。像弗雷德里希·埃伯特这类的居住区虽与理性主义的理论密切相关，但从各方面来看，这些居住区的实际形象却难以与"Siedlung"所包含的观念相一致。

我们至此已经考虑了"Siedlung"本身，而没有涉及（确切地说是忽视）其所产生的背景；只有参照20世纪20年代的柏林地区规划，我们才能分析城市中的居住区问题，这个问题实际上就是20世纪20年代柏林住房的问题。这个规划的基础是什么？它与某些最新模式的关系要比人们所想像的要接近得多。从总体上看，住房选择与地点没有多少关系。住房本身表现为城市体系中的一个要素，取决于体现城市脉搏的交通系统的发展。通过分区制，它促使政府和管理部门在市中心自成一区，而娱乐活动和体育设施等中心却被挤到了边缘地区。

这种模式甚至在今天还是一种基本的参照体系，尤其在那些居住区界线较为明确的地方。在柏林地区的规划中，我们可以看到以下一些情况：

1. 在城市中，居住区（Siedlung）并没有被设计成由不同部分组成的自主区域，这种实际状况因而比自主区域的设想要温和得多；

2. 德国理性主义者们已经认识到大城市及其形象的问题，人们只要想想为弗雷德里希大街设计的不同方案，尤其是密斯和陶特的方案，就可以知道这一点；

3. 解决柏林住房问题的方法，并没有完全不同于当时的基本住房模式，而是表现了新与旧的结合，这是很有意义的。

《城市建筑学》PP72-82
阿尔多·罗西著,黄士钧译
中国建筑工业出版社,2006年9月第1版

administrative district, while the centers for leisure activities, sports facilities, and the like were pushed to outlying areas.

This model is a basic reference even today, especially where the residential district is a more or less defined zone. Thus in the plan for Greater Berlin we find the following:

1. that the *Siedlungen* were not planned as autonomous district s within a city made up of different sectors - a formulation of this type would have been more revolutionary than was the reality;

2. that the German Rationalists in fact recognized the problem of the large city and its metropolitan image - one has only to think of the various projects for the Friedrichstrasse, in particular those of Mies van der Rohe and Bruno Taut;

3. that the solution to the housing problem in Berlin was not entirely different from the fundamental models of housing up to that time, but represented as well a synthesis of the new and the old, which is certainly a significant fact.

The Architecture of the City, PP72-82
Aldo Rossi
Ninth printing, 1997
The MIT Press

封闭社区

This chapter describes some of the findings of a two-year study of gated communities conducted during 1994 and 1995. The study involved a survey of representatives from gated community association boards, dozens of site visits, and interviews with focus groups and individual informants in nearly a dozen gated communities in the San Francisco Bay Area, Dallas/Fort Worth, and Miami/Fort Lauderdale.

Spatial Security
Gated communities in the United States go directly back to the era of the robber barons, when the very richest built private streets to seal themselves off from the hoi polloi. Later, during the 20th century, members of the East Coast and Hollywood aristocracies built more gated, fenced compounds. These early gated preserves were very different from the gated subdivisions of today. They were uncommon places for uncommon people. Now, however, the merely affluent and even many of the middle class can also have barriers between themselves and the rest of us. The first gates available to the mass market were those around master-planned retirement developments of the late 1960s and 1970s. Gates soon spread to resorts and country club communities, and then to middle-class suburban subdivisions. They have increased dramatically in number and extent since the early 1980s, becoming increasingly ubiquitous in most urban areas in the nation.

Gates range from elaborate two-story guardhouses manned 24 hours a day to roll-back wrought iron gates to simple electronic arms. Guardhouses are usually built with one lane for guests and visitors and a second lane for residents, who may open the gates with an electronic card, a punched-in code, or a remote control. Some gates with round-the-clock security require all cars to pass the guard, and management issues identification stickers for residents' cars. Others use video cameras to record the license plates and sometimes the faces of all who pass through. Unmanned entrances have intercom systems, some with video monitors, for visitors seeking entrance.

These security mechanisms are intended to do more than just deter crime. Both developers and residents

这一章描述了1994和1995年期间所做的关于封闭社区的两年研究的发现。这个研究牵涉到一次对封闭社区协会委员会代表的调查，几十次的场地访问，对来自于十多个位于旧金山湾区、达拉斯/沃斯堡、迈阿密/劳德代尔堡的封闭社区的焦点小组和个别信息源的访问。

空间的安全

封闭社区在美国可追溯到强盗大亨的时代，那时超级富有的人建私家街道以与老百姓隔离。之后，在20世纪，东岸和好莱坞的权贵建造更多的门控和围栏围合的院子。这些早期的门控保留地和今天的门控分支很不一样。那些是为非普通人建造的非普通场所。但是现在，刚刚富裕起来的、甚至中产阶级也能在他们之间或和其他人群之间建立屏障。最早面向大众市场的门控是用于20世纪60年代末和20世纪70年代的经过总体规划的退休社区开发。门控封闭形式很快蔓延到度假胜地和乡村俱乐部社区，然后到中产阶级郊区。它们自20世纪80年代开始在数量和范围上急剧增长，逐渐在全国的大部分都市地区无处不在。

社区大门从精心设计的24小时值班的两层警卫楼到铸铁大门到简单的电子手臂。警卫楼边上通常会有两条车道，一条给访客用，第二条给居民用，居民可以用电子卡片、代码打卡或遥控的方式来开门。有些全天值班的大门会要求所有车辆通过警卫，物业管理会给居民的车提供车贴以进行区分。其他的还有使用摄像机来记录车牌，有时甚至是记录所有经过的人脸。无人值守的入口有给访客寻找入口的对讲系统，有些会带视频监控。

这些安保机制的目的不仅仅是制止犯罪。开

发商和居民不只是将安保看作是用来免于犯罪，更是用来免于募捐者、推销员、淘气的青少年以及任何种类的陌生人的滋扰，不管他们是否有恶意。大门提供了一个将外来者排除在外的被庇护的公共空间。尤其对于高端的封闭社区的居民而言，他们已经居住在一个低犯罪环境当中，受限访问带来的私密和便利远比犯罪保护来的重要。

封闭社区的当代形式首先出现在阳光地带，在那里一直还流行。但现在已遍布全国，从西岸到东岸各州。因为它们主要是大都市群的现象，在大部分乡村地区如偏远的南部和大部分新英格兰地区还是很少见。

圈围的世界

对犯罪的恐惧已经成为影响我们日常生活几乎每个方面的重要因素。除了不断呼吁投入更多的公共资金和新的公共举措以打击犯罪，私营部门在预防和控制犯罪方面的角色正在蓬勃发展。封闭社区只是这一趋势的一部分。国家司法研究所的研究发现，现在就职于安保领域的人，从设备制造商到运钞的车司机，三倍于被官方执法机构雇佣的人。私营安保机构比公共执法部门多花费73%，现在显然是全国主要的安保资源。

媒体在整个国家范围的渗透以及它们对于戏剧化的人性故事所展现的贪得无厌的胃口使得一个西北地区小镇的罪案报道会覆盖从西雅图到迈阿密的广大地区。这个推动力量加注了对犯罪的恐惧和犯罪状况正在恶化的固执观感——即使犯罪率在1990年代早期实际上是下降的。几乎有90%的美国人认为犯罪状况更为恶化，但在1981到1989年间的城市暴力犯罪率降低了25%。而且尽管55%的民众担心成为犯罪受害者并且相同百分比的民众感觉警察保护不够，但当被问到社区当中什么最困扰他们的时候，只有7.4%的人提到了犯罪。

犯罪现象表面上看来的随机性也是这种恐惧感增高的原因。城市被认为是犯罪的核心地区，没人可以保证是他们是安全的。青年被等同于犯罪，少数族裔的青年为这种日益升高的恐惧感承受了不成比例的负担。陌生人不管如何被描述，就是恐惧和不信任的自动诱因。对于将自己关在社会信任、车流等于陌生人、陌生人是坏的、而坏人意味着犯罪这一新的等式中的许多社区而言，车流形成一个等重的甚至是更大担忧的原因。

从现实来看，犯罪问题对低收入人群而言远比生活较好的人群要来的严重。国家司法统计局

view security as not just freedom from crime, but also as freedom from such annoyances as solicitors and canvassers, mischievous teenagers, and strangers of any kind, malicious or not. The gates provide a sheltered common space that excludes outsiders. Especially to the residents of upper-end gated communities, who can already afford to live in very-low-crime environments, the privacy and convenience of controlled access are more important than protection from crime.

Gated communities in their contemporary form emerged first in the Sunbelt, and they remain most common there. But they are now found across the country, in states from the West Coast to the East. Because they are primarily a phenomenon of metropolitan agglomerations, they are rarities in largely rural areas such as the deep South and most of New England.

A Walled World

Fear of crime has become an influential factor in nearly every aspect of our daily lives. In addition to the constant calls for more public monies and new public initiatives to combat crime, the private sector's role in crime prevention and control is booming. Gated communities are only one part of this trend. A National Institute of Justice study found that three times as many people now work in the security field, from equipment manufacturers to armored car drivers, as are employed by official law enforcement agencies. The number of security guards has doubled in the last decade and now surpasses the number of police. Private security outspends public law enforcement by 73 percent and is now clearly the nation's primary protective resource.

The national reach of the media and their insatiable appetite for dramatic human interest stories mean that a crime committed in a small northwestern town is reported from Seattle to Miami. This dynamic fuels the fear of crime and the dogged perception that crime is worsening - even in periods like the early 1990s, when crime rates actually dropped. Almost 90 percent of Americans think crime has gotten worse, but the violent crime rate in cities dropped 25 percent between 1981 and 1989. And although 55 percent worry about being a victim of crime and the same percent feel inadequately protected by the police, only 7.4 percent mention crime when asked what bothers them in their neighborhoods.

The seeming randomness of crime is also responsible for this heightened fear. Cities are viewed as the core area of crime, but no one can be certain they are safe. Youth and crime are now synonymous, and minority youth bear a disproportionate burden of this rising fear. Strangers of any description are an automatic inducement to fear and distrust. This is one reason that traffic is of equal or even greater concern to many neighborhoods that close themselves off in the new equation of social trust, traffic equals strangers, strangers are bad, and bad means crime.

Realistically, crime is a far greater problem for lower-income people than for the better off. Data from the Bureau of Justice Statistics's *National Crime Victimization Survey* show that it is also a greater problem in cities than in suburbs or rural areas. The rates for both violent crime and household crime such as burglary are about 35 percent lower in the suburbs than in cities. City residents are one and a half times more likely than suburbanites to be a victim of a violent crime or a household burglary. Yet gates are primarily a suburban phenomenon. The real danger of crime bears no necessary relationship to the fear of crime. In places with high crime rates, places with low crime rates, places where crime is rising, and places where crime is dropping, fear can spur the gating of neighborhoods that were once open to their surroundings.

The results of the survey of homeowner association boards in gated communities show that security is a primary concern for those who buy in gated communities. The respondents certainly thought that they and their neighbors were drawn to fortifications around their subdivisions; nearly 70 percent of respondents indicated that security was a very important issue in the ultimate decision of residents to live in their gated communities. Only 1 percent thought that security was not an important motivation.

Gated communities are a response to the rising tide of fear. They can be classified in three main categories. First are "lifestyle communities," where the gates provide security and separation for the leisure activities and amenities within. Subtypes within this category include retirement communities; golf, country club, and resort developments; and new towns. Second are "prestige communities," where the gates symbolize distinction and prestige and attempt to create and protect a secure place on the social ladder. Subtypes include enclaves for the rich and famous; developments for senior executives and managers, and successful professionals; and executive subdivisions. And third are "security zones," where community safety is the primary goal. They may be center city or suburban, in rich or poor areas, but gates are primarily a protection from some threat, real or perceived. In the first two categories, the developer builds gates as an amenity and image that helps sell houses; in the security zone, it is the residents who build gates, retrofitting their neighborhoods to shield them from the outside world.

…

Gates as Crime Prevention

Residents say repeatedly that they want to protect themselves from crime, reduce traffic, and control their community. And they believe that the gates work. As one developer said, "Gated communities weren't around a while back. The world is a drastically different place as a result of increased violence and decreased municipal services." The gates, he believes, create "a friendlier place, an open community because of the perception of safety, insularity, and being in their own little bubble." In the authors' survey of gated communities, over two-thirds of respondents believed that their community experienced less crime than the surrounding area. Of this amount, a full 80 percent attributed the difference to

的全国犯罪受害者调查报告的数据显示犯罪在城市比在市郊或乡村地区要更成问题。暴力犯罪和家庭犯罪率，如入室盗窃，郊区要比城市低35%。城市居民要比郊区居民多一倍半的可能成为暴力犯罪或入室盗窃的受害者。但封闭却只是一种郊区现象。犯罪的真实危险和对于犯罪的恐惧没有必然的联系。高犯罪率地区、低犯罪率地区、犯罪率升高地区、犯罪率降低地区，恐惧都将激发曾经对周围开放的社区进行封闭。

封闭社区房主协会委员会的调查结果表明，安保是在封闭社区购房者的首要考虑。受访者认为他们和他们的邻居是被这个他们周围的安全堡垒吸引来的；接近70%的受访者指出安保是居民做出居住于封闭社区这一最后决定的非常重要的因素。只有1%的人认为安保不是一个非常重要的动机。

封闭社区是对日益增高的恐惧浪潮的回应。它们可以分为三大类。首先是"生活方式社区"，封闭为内部的休闲活动和设施提供了安保和隔离。这个类别中的亚类包括退休社区；高尔夫、乡村俱乐部以及度假村；还有新城发展。其次是"声望社区"，封闭在这里象征着荣誉和声望，并试图创造和保护社会等级上的一个安全之地。其亚类包括富人和名人的飞地；高管、经理人、成功职业人士的社区；还有普通管理人员的社区。第三类是"安保区"，在那里社区安全是首要目标。它们可能是中心城市或郊区，可能在富裕或贫困地区，但封闭主要是对某种威胁的防护，不管是真实的还是感觉的。在前面两个类型中，开发商建立门控封闭作为帮助销售房屋的一种附加设施和广告形象；而在安保区，是居民们自己建立门控封闭，改造他们的社区以保护他们不受外部世界的侵害。

……

封闭作为犯罪预防

居民们反复说他们想要保护自己免受犯罪、降低车流、并管控他们的社区。他们也坚信封闭确有成效。正如一位开发商说，"封闭社区不久前还没有，但这个世界因为日益增加的犯罪和减少的市政服务而变成一个截然不同的地方。"封闭，他认为创造了"一个更为友善的场所，一个因为感受到安全、孤立、以及因禁在他们自己小泡泡内而开放的社区。"在作者对封闭社区的调查中，超过三分之二的受访者相信他们的社区比周围地区经历更少的犯罪活动。在一数字中的80%，将这个差异归功于封闭。

但实际情况如何呢？所有这类安保对于犯罪有实际影响吗？证据并不能证实这一点。所有作者进行焦点小组研究地区的警察报告在封闭和非封闭社区之间最多也就是细微的差别。大部分警察发现没有区别；犯罪率因地区而不同而不在于同一个地区的封闭和非封闭社区之间的区别。有些甚至认为封闭阻碍了警察的作用，因为封闭延缓了响应时间，围墙阻挡了视线，居民有一种虚假的安全感，使得他们车库门大开，门窗也不上锁。

在封闭安装门控之前和之后都有犯罪率数据的地方，犯罪预防的证据是模糊不清的，甚至在安保区域范围内。许多可用的数据质量很差，但甚至是最可靠的研究也只是表明封闭的结果是混合的或并不显著。例如，在20世纪70年代中期由一个为政府和社区提供封闭社区设计的规划师所作的对圣路易斯的封闭街区和类似的开放街区的对比研究中，发现犯罪发生率有很大的不同，但是发现封闭街区通常显示更低的发生率，至少对于某些类型的犯罪如是。开放和封闭社区的最大差别在于观感：那些门控后面的人群在街道上感觉更为安全。

劳德代尔堡的警方进行了两项更彻底和更广泛的研究。第一份研究发现在封闭街区暴力和财产犯罪的发生率并无显著变化。汽车盗窃、入室盗窃、和其他一些犯罪在实施封闭后确大幅降低，但并不能持续长时间。第二份研究对比了几个封闭街区和整个城市之间的犯罪率的变化，得出结论门控和路障对此无显著影响。研究还包括了一份巡警的调查报告，显示他们大多数不喜欢街区封闭。多数人认为它们不减少犯罪，但反而延缓了响应时间并抑制了警察巡逻。

门控和路障作为控制犯罪的手段时的含糊的、零星的成功和失败，显示其尽管让人们觉得更安全，但可能并不真的更为安全。所以，恐惧和焦虑以自身为食。门控和围墙反映了恐惧，起着观察另外一边所产生的危险的日常提醒器的作用，但同时无助于改善现实。当然，在当下封闭的潮流之后，并不只是出于对犯罪的恐惧。面对经济、人口、和社会变化带来的日益增高的焦虑，封闭令人安心。它们将人们觉得脆弱的世界排除在外。即使犯罪在封闭社区内部降低了，外面城市和郊区的街道并无变化，大都会也并无变化。一些封闭社区的支持者认为通过提供私营安保，这些社区减轻了公共警力的负担，可以释放这些

the gates.

But what is the reality? Does all of this security have any real impact on crime? The evidence does not suggest that it does. Police in all the areas where the authors conducted focus groups reported at best marginal differences in crime between gated and ungated developments. Most found no difference; crime rates varied by area but not between gated and ungated neighborhoods in the same area. A few even believed they hampered police efforts, because gates slowed response time, walls blocked sight lines, and residents gained a false sense of security, leading them to leave garage doors open and doors and windows unlocked.

Evidence of crime prevention is ambiguous, even in security zone communities, where data on crime rates are available for both before and after gating or barricading. Much of the available data are of poor quality, but even reliable studies show mixed or slight results from gating. For example, a study of closed-street neighborhoods in St. Louis compared with similar open-street neighborhoods in the mid-1970s by a planner who consults with governments and communities to design gating plans found great variations in the incidence of crime, but found that the closed-street neighborhoods in general showed lower rates, at least for some types of crime. The biggest difference between open- and closed-street neighborhoods was in perception: those behind gates felt much safer on their streets.

Two of the more thorough and wide-ranging studies were conducted by police in Fort Lauderdale. The first found no significant change in rates for violent or property crime in a closed-street neighborhood. Auto theft, burglary, and some other crimes dropped considerably immediately after closure, but none were sustained for more than a short time. A second study compared the change in crime rates in several closed-street neighborhoods with that of the city as a whole and concluded that the gates and barricades had no significant effect. That study also included a survey of patrol officers and found that the majority disliked the street closures. Most thought that they do not reduce crime, but do slow response time and inhibit police patrols.

The ambiguous and spotty successes and failures of gates and barricades as measures to control crime indicate that although people may feel safer, they probably are not significantly safer. Thus, fear and anxiety feed on themselves. Gates and walls reflect fear and serve as daily reminders of the perceived dangers on the other side at the same time they do little to improve the reality. More than the fear of crime, of course, is behind the current wave of gating. Gates are reassuring in the face of anxiety heightened by economic, demographic, and social change. They exclude a world where one feels vulnerable. Even if crime were reduced in the gated developments, the city or suburban streets outside are unchanged and the metropolitan area is unchanged. Some proponents of gated communities argue that by providing private security, these developments are relieving the public policing burden, freeing resources to be used elsewhere. In most cases, however, they augment rather than replace police services, especially where residential street patrols are not a significant part of police activities, as in the low-crime suburbs where gated communities are most common.

You Can Run But You Can't Hide
There is little doubt that urban problems are the stimuli for this wave of gating. A growing underclass, high levels of foreign immigration, and a restructured economy are leaving many feeling insecure. Gated communities are a search for stability and control in the face of these dramatic demographic changes. The drive for separation, exclusion, and protection that gated communities represent is just a part of the larger spatial pattern of segmentation in which this country is increasingly separated by income, race, and economic opportunity.

Economic segregation is scarcely new. In fact, zoning and city planning were designed in part to preserve the position of the privileged by subtle variances in building and density codes. But the gated communities go farther in several respects. They create physical barriers to access, and they make public, not merely individual, space private. Many of these communities also privatize previous public responsibilities, such as police protection, parks and recreation, and a range of mundane civic functions from trash collection to street maintenance, leaving the poor and less well-to-do dependent on the ever-reduced services of city and county governments.

This privatization - in both senses is one of the more serious effects of gated communities on social equity and the broader community. The new developments create a private world that need share little with its neighbors or the larger political system. This fragmentation undermines the very concept of organized community life. We no longer speak of citizens, but of taxpayers, who take no active role in governance but merely exchange money for services. In the gated communities, many say they are taking care of themselves and have no desire to contribute to the common pool serving their neighbors in the rest of the city. In areas where gated communities are the norm, not the exception, this perspective has potentially severe impacts on the common welfare. Failed cities and gated communities are a dramatic manifestation of the fortress mentality growing in America.

Within metropolitan areas, poverty and economic inequality are no longer limited to the inner cities. Even formerly well-established, "good" suburbs now have their share of social and structural problems. The suburbs are becoming urbanized, so that many might now be called "outer cities," places with many problems and pathologies traditionally thought to be restricted to big cities.

Gated and barricaded communities are themselves a microcosm of the larger spatial pattern of segmentation

资源用于其他地方。在大多数情况下，它们是增强而不是替换警察服务，特别是当居住区街道巡逻并不是警察活动的重要部分时，就像在封闭社区最为常见的低犯罪率的郊区。

你能逃跑但你没地方躲藏

很少有人怀疑城市问题是这一波封闭潮流的刺激因素。日益增多的下层阶级，高水平的外国移民，和转型的经济，使得许多人感到不安。封闭社区是在面对这个急剧人口变化的对于稳定和管控的探索。封闭社区所代表的隔离、排斥、和保护的驱动力，只是这个国家被收入、种族、经济计划日益隔离的、更大的被分割的空间模式的一部分。

经济隔离并不是新鲜事物。实际上，区划和城市规划部分就是旨在通过建筑和密度的设计规范中微妙的差异来保证特权阶级的地位。但封闭社区在几个方面走得更远。它们制造访问的物理障碍，它们将公共空间，而不只是个人的，转变为私人空间。许多这类社区将曾经的公共义务私营化，比如警察保护、公园和休憩、以及从垃圾收集到街道维护的一系列世俗的市政功能，而将穷人和不那么富裕的人丢给市或县政府一直被缩减的服务。

这类私营化——在两方面都是封闭社区在社会平等和更广泛社区方面的严重影响之一。新的房地产开发创造了一个不需要和邻居或更大的政治系统共享的私人世界。这种碎片化破坏了有组织社区生活的概念。我们不再提及公民，而只是说纳税人，其并不积极参与治理，只是以金钱换取服务。在封闭社区，许多人说他们照顾自己，没有意愿为服务城市其他地区邻居的公共储备作出贡献。在封闭社区已成为规范而非例外的地区，这个观点对于共同福利具有潜在的严重冲击。失效的城市和封闭社区是在美国堡垒心态增长的激烈表现。

在大都会地区，贫困和经济不平等不再局限于内部城市。即便是曾经历史悠久的"好"郊区，现在有了它们的社会和结构问题。郊区经历着城市化，以至于它们中的许多现在可以被称为"外城"，一个拥有许多曾经只可能属于大城市的问题和病理的地方。

门控和路障社区本身就是一个更大的分割和隔离的空间模式的缩影。城市和郊区之间、富人和穷人之间日益增长的分歧正在创造一个新的模式，增强孤立和排斥在有利于一些人的同时却强加给另一些人的代价。这些"地盘争夺战",大部分以封闭社区体现,是土地利用规划的一股令人不安的趋势。当公民们将自己划分到同质的、独立的单元中,他们和更大的政体和社会间的纽带减弱了,增加了解决市政问题的努力的阻碍,遑论区域问题了。正如黑鹰乡村俱乐部封闭社区的一位居民在焦点小组所说,"人们厌倦了政府处理问题的方式。因为你不能真正控制如何花钱,你觉得公民权被剥夺了。如果法庭继续释放罪犯,而我们不对罪犯起诉,并以我们现在花钱的方式继续花钱,那么至少在黑鹰,我还能对自己如何生活有一点控制权。"

这一建设堡垒的现象对政策有巨大影响。允许一些公民脱离公众联系,并将其他人排除于他们的经济和社会利益之外,直接瞄准了这个国家的社区和公民权的概念基础。老的社区流动和相互责任的观点被这些新的社区模式松绑。当邻里之间的分隔需要武装巡逻和电篱来阻挡其他公民时,如何衡量这个国家的政体呢?当公共服务、甚至地方政府被私营化之后,当社区责任止步于社区门口时,社会和政治民主的功能及其概念会变成怎样?简而概之,这个国家能在没有社会接触的前提下完成它的社会契约吗?

《通过房地产开发和管理减少犯罪》PP53-70
"分离之地:封闭社区的犯罪和安保"
Edward J. Blakely and Mary G. Snyder
此段卓旻译

and separation. The growing divisions between city and suburb and between rich and poor are creating new patterns that reinforce the costs that isolation and exclusion impose on some at the same time they benefit others. These "turf wars," while most dramatically manifested by gated communities, are a troubling trend for land use planning. As citizens separate themselves into homogenous, independent cells, their ties to the greater polity and society become attenuated, increasing resistance to efforts to resolve municipal, let alone regional, problems. As one resident of the gated country club development Blackhawk said in a focus group, "People are tired of the way the government has managed issues. Because you don't really have control over how the money is spent, [you] feel disenfranchised. If the courts are going to release criminals and we're going to continue not to prosecute people and continue to spend money the way we've been spending it, and I can't change it, at least here in Blackhawk, I have a little control over how I live my life." This phenomenon of building fortresses has enormous consequences for policy. Allowing some citizens to secede from public contact and to exclude others from their economic and social privilege aims directly at the conceptual base of community and citizenship in this country. The old notions of community mobility and mutual responsibility are loosened by these new community patterns. What is the measure of nationhood when the divisions between neighborhoods require armed patrols and electric fences to keep out other citizens? When public services and even local governments are privatized, when the community of responsibility stops at the subdivision's gates, what happens to the function and the very idea of a social and political democracy? In short, can this nation fulfill its social contract without social contact?

Reducing Crime through Real Estate Development and Management, PP53-70
"Separate Places: Crime and Security in Gated Communities."
Edward J. Blakely and Mary G. Snyder

拥挤文化

ATHENS
As a boy - "where the bias, they say, is set" - Hugh Ferriss is given a picture of the Parthenon for his birthday.
It become his first paradigm. "The building seemed to be built of stone. Its columns seemed to be designed to support a roof. It looked like some sort of temple… I learned in due time that all those impressions were true…"
"It was an honest building," built "in one of those fortunate periods when engineers and artists worked enthusiastically together and when the populace warmly appreciated and applauded their alliance…"
The image of the Parthenon inspires Ferriss to become an architect; when he gets his degree he leaves his hometown, St. Louis, for Manhattan. For him, New York represents a new Athens, the only possible birthplace of new Parthenons.
"One wanted to get to the Metropolis. In New York… an indigenous American architecture would be in the making, with engineers and artists working enthusiastically together - and maybe even with the populace warmly appreciating and applauding their alliance…" But at his first job at Cass Gilbert's office - then designing the Woolworth Building - Ferriss' "juvenile enthusiasm is in for a jolt." The contemporary architecture of Manhattan does not consist of the production of new Parthenons but of the pilferage of all useful elements of past "Parthenons," which are then reassembled and wrapped around steel skeletons.
Instead of a new Athens, Ferriss finds ersatz antiquity. Instead of contributing to the design of dishonest buildings, Ferriss prefers the technical and strictly neutral role of renderer; he is made "delineator" in Gilbert's office.

PILOT
By the early twenties he has established himself as independent artist in his own studio. As a renderer Ferriss is the puritanical instrument of a coalition of permissive eclectics: the more convincing his work, the more he promotes the realization of proposals he dislikes.

雅典

当还是一个男孩——"就像他们说的偏见"——休·费里斯得到了一张帕台农神庙的照片作为他的生日礼物。

这成了他的第一种范式。"那个建筑似乎由石头建成，它的柱子似乎被设计成用于支撑屋顶。它看上去像是某种庙宇……没过多久我发现这些印象都是真的……"

"这是一个诚实的建筑，"建于"那几个幸运的年代之一，那时候工程师和艺术家热情合作，而人们也热烈地赞赏这种联盟并为之喝彩……"

当他拿到学位离开家乡圣路易斯而来到纽约时，帕台农神庙的形象鼓舞费里斯成为一名建筑师。对他来说，纽约代表着新雅典，是新的帕台农神庙可能诞生的唯一地方。

"人们想要来到大都会。在纽约……一种本土的美国建筑可能正在产生，工程师和艺术家热情合作——人们甚至也热烈地赞赏这种联盟并为之喝彩……"但他在凯斯·吉尔伯特工作室的第一份工作——那时在设计伍尔沃斯大楼——费里斯的"年轻人的热情进入动荡期"。曼哈顿的当代建筑不在于生产新的帕台农神庙，而是从过去的"帕台农神庙"中窃取一切有用的元素再将之重新组装和包裹到钢铁骨架上。

不再是新雅典，费里斯发现了替代的古迹。不再是为不诚实的建筑设计出力，费里斯更喜欢作为一位渲染师那样的技术性的和严格中性的角色；他在吉尔伯特工作室担任了"绘图师"。

导航员

在20世纪20年代初，他已经将自己定位于

不同的观点 · 27

在自己工作室创作的独立艺术家。作为一位渲染师，费里斯是一个宽泛的折衷主义联盟的及其严谨的工具：他的工作越令人信服，他也就更多地在推动他所不喜欢的方案的实施。

但是费里斯发现了一种逃离这个两难境地的方法：一种将他的意图和他的客户们的意图隔离开来的技法。他用木炭条，一种印象主义的介质，依赖于平面的暗示和本质上对于涂抹的操纵。

通过使用这种不能描绘折衷性的、占据曼哈顿建筑师心灵的表面琐事的介质，费里斯的画作所剥夺的和所给予的一样多。通过每一次的再现，他可以从过度的表皮下解放出一个"诚实的"建筑。

费里斯的描绘，即使是为了曼哈顿的建筑师们吸引客户——通过客户以吸引更大的人群——却是对他们假装想要表现的项目的批判，是对于他们所基于的反动蓝图的争论性的"校正"。曼哈顿的建筑师们在对费里斯的服务产生依赖这一点上是共通的，这加强了对这些被校正项目的累积影响。它们凝聚成一个未来曼哈顿的一致的愿景。那个愿景在曼哈顿的居民中变得越来越受欢迎——以至于费里斯的绘画独自就代表了曼哈顿的建筑，而根本没人在乎设计每个项目的个体建筑师。在一种经过计算的模糊性当中，费里斯的图像恰恰创造了那部分"热烈赞赏和喝彩的"观众，这是在他青年时代被他认为是新雅典产生的条件之一。通过成为一个帮助者，这个伟大的绘图师转变成一个领袖。

"他给最没有希望的构图注入诗歌般的角度……利用他的天赋的最好的办法就是扔掉方案上床睡觉，然后第二天早上会发现整个设计已经完成了。他是一个完美的自动导航员。"

研究

在他的商业工作的同时，费里斯和几个进步建筑师，如雷蒙德·霍德和哈维·威里·科贝特，一起调查了曼哈顿的真实问题。这个研究专注于1916年区划法规的未开发潜力，以及它所用于描述每个曼哈顿街区的理论包膜。

费里斯的图纸是对无限变量的第一层揭示——既是形式的也是心理的——包含在它的基本的合法形状之中。在穷尽个体类别之后，他创作了这些个体变量所形成组合体的第一个具体的意象：作为曼哈顿最终命运的巨型村落。

对于费里斯而言，这个以不能被触摸形式存在的城市是新的雅典："当一个人思考这些形状

But Ferriss discovers an escape from this dilemma: a technique that isolates his own intentions from those of his clients. He draws in charcoal, an imprecise, impressionistic medium that relies on the suggestiveness of planes and the manipulation of what are, essentially, smudges.

By using the one medium incapable of depicting the eclectic surface trivia that preoccupy Manhattan's architects, Ferriss' drawings strip as much as render. With each representation he liberates an "honest" building from under the surface excess.

Ferriss' delineations, even as they are intended to seduce clients for Manhattan's architects - and through them, the larger population - are critiques of the projects that they pretend to embody, polemical "corrections" of the reactionary blueprints on which they are based. That Manhattan's architects have only their dependence on Ferriss' services in common reinforces the cumulative impact of these corrected projects. They coalesce into a coherent vision of a future Manhattan.

That vision becomes increasingly popular with Manhattan's inhabitants - to the point where Ferriss' drawings alone represent Manhattan's architecture, regardless of the individual architect who designed each project. In their calculated vagueness Ferriss' images create exactly that "warmly appreciative and applauding" audience that he has identified in his youth as the condition for the birth of a new Athens. From being a helper, the great delineator becomes a leader.

"He can pump perspective poetry into the most unpromising composition… The best way to utilize his talent would be to toss over the plans, go to bed and turn up the next morning to find the design all done." "He is the perfect *automatic pilot*…"

RESEARCH

Simultaneously with his commercial work, Ferriss investigates the true issues of Manhattanism with several progressive architects, such as Raymond Hood and Harvey Wiley Corbett. This research centers on the unexplored potential of the 1916 Zoning Law and the theoretical envelope it describes on each Manhattan block.

Ferriss' drawings are the first revelations of the infinite variations - both formal and psychological - contained within its basic legal shape. After exhausting the individual categories, he produces the first concrete image of their final assembly: the Mega-Village that is Manhattan's final destiny.

For Ferriss, this new city of untouched forms is the real new Athens: "As one contemplates these shapes, images may begin to form in the mind of novel types of building which are no longer a compilation of items of familiar styles but are, simply, the subtleizing of these crude masses…"

That city would establish him, Ferriss the renderer, as its chief architect, since its drastic nakedness is what his charcoal medium has always anticipated.

Signs of its imminence are already in the air, baffling the traditional architects; "conservative architectural standards were thrown into confusion. At point after point designers found themselves faced with restrictions which made the erecting of familiar forms impossible…"

LABORS
In 1929 Ferriss publishes the summation of his labors, *The Metropolis of Tomorrow*.
The book is divided into three parts: *Cities of Today*, a collection of his renderings for other architects; Projected Trends, his variations on the theme of the 1916 law; and *An Imaginary Metropolis*, Ferriss' new Athens.
There are 50 drawings in the book, each "explained" by a text that is the verbal equivalent of the drawings' charcoal vagueness.
The structure of the book is modeled on the clearing of a persistent fog bank: from "it is again dawn, with an early mist enveloping the scene," via "as the mist begins to disperse," to "a little later, the general clearing of the air allows us to check upon our first impressions…"
This "plot line" corresponds to the three parts of the book: an imperfect past - the work of the other architects; a promising present - the annunciation and theoretical elaboration of the Mega-Village of the 1916 Zoning Law; and the shining future of Ferriss' imaginary Metropolis, which is one version of that village: "a wide plain, not lacking in vegetation, from which rise, at considerable intervals, towering mountain peaks…"

WOMB
But actually, the divisions are less important than the continuity of all these quasi-nocturnal images. The genius of Ferriss' production is in the medium of his renderings itself, the creation of an artificial night that leaves all architectural incidents vague and ambiguous in a mist of charcoal particles that thickens or thins whenever necessary.
Ferriss' most important contribution to the theory of Manhattan is exactly the creation of an illuminated night inside a cosmic container, the murky *Ferrissian Void*: a pitch black architectural womb that gives birth to the consecutive stages of the Skyscraper in a sequence of sometimes overlapping pregnancies, and that promises to generate ever-new ones.
Each of Ferriss' drawings records a moment of that never-ending gestation. The promiscuity of the Ferrissian womb blurs the issue of paternity.
The womb absorbs multiple impregnation by any number of alien and foreign influences - Expressionism, Futurism, Constructivism, Surrealism, even Functionalism - all are effortlessly accommodated in the expanding receptacle of Ferriss' vision.
Manhattanism is conceived in Ferriss' womb.

…

VENICE
A hundred profound solitudes together constitute the city of Venice. That is its charm. A model for the men of the future.
- Friedrich Nietzsche

时，意象开始形成于新颖建筑类型的思维当中，这不再是熟悉风格的汇编，而是纯粹对于这些粗略体量的精雕……"

那个城市将任命他——绘图师费里斯，为总建筑师，因为城市激烈的裸露性正是他的木炭条介质所一直期待的。

城市的紧迫性迹象已经能在空气中嗅到，这使得传统建筑师困惑；"保守的建筑标准陷入混乱。设计师们发现他们面临限制，将使得竖立起熟悉的形式不再可能……"

成果

1929年费里斯出版了他的成果总结，明天的大都会。

这本书分为三个部分：城市的今天，即他为其他建筑师所绘的效果图的选集；预测的趋势，他以1916年法规为主题的变异；想象中的大都会，也即费里斯的新雅典。

书中有50幅画作，每幅作品以相当于画作上的炭条所表现出的模糊性的语言等价物的文本进行"解释"。

这本书的结构仿照对于永久性雾障的清除模式："又是黎明，晨雾包围现场"，然后"雾开始消散"，到"过一会儿大部分的雾消了，我们可以检视一下我们的第一印象……"

这个"情节线索"对应书中的三个部分：一段不完美的过去——其他建筑师的工作；一个有希望的现在——对于1916年区划法规中的巨型村落的报喜和理论阐述；费里斯幻想中的大都会的闪耀未来，那是那个村落的一个版本。"一块不缺少植被的平坦平原，有着巨大间隔的高耸山峰拔地而起……"

子宫

但实际上，每个分部并不比所有这些半黑夜图像的连续性更重要。费里斯的创作的聪明之处在于他的绘画自身的介质，在必要时可随时可加厚或变薄的碳粉中，让所有建筑事件变得模糊和模棱两可的人造黑夜的创造。

费里斯对他的曼哈顿理论的最重要贡献正是在宇宙容器中的发光的黑夜的创造，阴沉的费里斯式的虚空：一个漆黑一片的建筑子宫，它以有时候交叠孕育的序列诞生出摩天大楼的连续阶段，它承诺会一直诞生出新的东西。

每幅费里斯的图都记录下那永无休止的妊娠的时刻，费里斯式子宫的滥交模糊了亲子关系。

这个子宫吸收着任意数量的异族的和外国的影响所带来的受孕——表现主义、未来主义、构成主义、超现实主义、甚至功能主义——所有这些毫不费力地被容纳在代表费里斯愿景的扩展容器中。

曼哈顿主义是在费里斯的子宫中孕育的。

……

威尼斯

一百个极度孤独的荒岛构成了水城威尼斯。那是它的魅力所在。那是一个给未来人类的模型。
——弗雷德里希·尼采

但是纽约，再加上其他很多东西，恰似威尼斯正在形成。所以这些用于城市缓慢向前成型的丑陋的用具，所有这些讨人嫌的过程，不管是物理的和化学的、结构的和商业的，都必须被认可、被表达出来、并借助诗歌的视野来制成城市的美丽和浪漫的一部分。
——J 门罗·休利特，主席，纽约建筑联盟
纽约：国家的大都会

最精确和确实的解决拥挤问题的提案来自于哈维·威里·科贝特，一个杰出的关注曼哈顿和摩天大楼的思想家，也是在哥伦比亚大学为年轻一代授课的教师。

在他的升高的拱廊人行道方案中（最早于1923 年提出），整个城市的地面——现在是各种交通模式构成的一片混乱——将逐渐让渡给单一的汽车交通。这个平面上开挖的壕沟还可以让快速交通更快地贯穿大都会。如果汽车还需要更多空间，那么现存大楼的边缘还可以后退来为交通提供更多的区域。在二层，从大楼侧面挖出拱廊人行道。拱廊在街道两边形成一个连续网络；天桥提供了这个连续性。沿着拱廊，店面和其他公共设施嵌入在大楼里。

通过这种分离，原来街道的容量至少增加 200 个百分点，如果这条街道占据更多的地面，那么容量增加的更大。

最终，根据科贝特的计算，整个城市的地面可以被用作单一交通，如同汽车的汪洋，那么交通潜力可以增加 700 个百分点。

"我们将看到这样一个城市，人行道有拱廊覆盖，在建筑物边界之内，但比现在的街面高一

But New York, in addition to being a lot of other things, is a Venice in the making, and all the ugly paraphernalia by means of which this making is slowly going forward, all the unlovely processes, physical and chemical, structural and commercial, must be recognized and expressed and by the light of poetic vision be made a part of its beauty and romance.
- J Monroe Hewlett, President, Architectural League of New York
New York: The Nation's Metropolis

The most precise and literal proposal to solve the problem of congestion comes from Harvey Wiley Corbett, prominent thinker about Manhattan and the skyscraper, and teacher of the younger generation at Columbia University.

In his scheme for elevated and arcaded walkways (first proposed in 1923), the entire ground plane of the city – now a chaos of all models of transportation – would gradually be surrendered solely to automotive traffic. Trenches in this plane would allow fast traffic to rush through the Metropolis even faster. If cars needed more room again, the edges of existing buildings could be set back to create still larger areas for circulation. On the second story pedestrians walk along arcades carved out of the buildings. The arcades form a continuous network on both sides of streets and avenues; bridges provide its continuity. Along the arcades, shops and other public facilities are embedded in the buildings.

Through this separation, the capacity of the original street is increased at least 200 percent, more if the road consumes still larger sections of the ground plane.

Ultimately, Corbett calculates, the entire surface of the city could be a single traffic plane, an ocean of cars, increasing the traffic potential 700 percent.

"We see a city of sidewalks, arcaded within the building lines, and one story above the present street grade. We see bridges at all corners, the width of the arcades and with solid railing. We see the smaller parks of the city (of which we trust there will be many more than at present) raised to this same side-walk arcade level… *and the whole aspect becomes that of a very modernized Venice, a city of arcades, plazas, and bridges, with canals for streets, only the canals will not be filled with real water but with freely flowing motor traffic, the sun glistering on the black tops of the cars and the buildings reflecting in this waving flood of rapidly rolling vehicles.*"

"From an architectural viewpoint, and in regard to form, decoration and proportion, the idea presents all the loveliness, and more, of Venice. There is nothing incongruous about it, nothing strange…"

Corbett's "solution" for New York's traffic problem is the most blatant case of disingenuity in Manhattanism's history. Pragmatism so distorted becomes pure poetry. Not for a moment does the theorist intend to relieve congestion; his true ambition is to escalate it to such intensity that it generates – as in a quantum leap – a completely new condition, where congestion becomes mysteriously positive.

Far from solving any problems, his proposal is a

metaphor that orders and interprets an otherwise incomprehensible Metropolis.
With this metaphor, many of Manhattan's latent themes are substantiated: in Corbett's "*very modernized Venice*" each block has become an island with its own lighthouse, the Ferrissian phantom "house". The population of Manhattan – journeying from block to block –would finally, and literally, inhabit a metropolitan archipelago of 2,028 islands of its own making.

CONGESTION
Ferriss, Corbett and the authors of the Regional Plan have invented a method to deal rationally with the fundamentally irrational. They know instinctively that it would be suicide to solve Manhattan's problems, that they exist by the grace of these problems, that it is their duty to make its problems, if anything, forever insurmountable, that the only solution for Manhattan is the extrapolation of its freakish history, that Manhattan is the city of the perpetual *flight forward*.
The Planning of these architects – assembled in the Regional Plan Committee – must be the opposite of objective. It consists of the imposition on the explosive substance of Manhattan of a series of metaphoric models – as primitive as they are efficient – that substitute for literal organization –impossible in any case – a form of poetic control. The "house" and "village" of the 1916 Zoning Law, Ferriss' "buildings like monuments" and finally Corbett's Manhattan as a "very modernized Venice" together form a deadly serious *matrix of frivolity*, a vocabulary of poetic formulas that replaces traditional objective planning in favor of a new discipline of *metaphoric planning* to deal with a metropolitan situation fundamentally beyond the quantifiable.
Congestion itself is the essential condition for realizing each of these metaphors in the reality of the Grid. Only congestion can generate the super-house, the Mega-Village, the Mountain and finally the modernized automotive Venice.
Together, these metaphors are the foundation of a *Culture of Congestion*, which is the real enterprise of Manhattan's architects.

CULTURE
The Culture of Congestion proposes the conquest of each block by a single structure.
Each Building will become a "house" – a private realm inflated to admit houseguests but not to the point of pretending universality in the spectrum of its offerings.
Each "house" will represent a different lifestyle and different ideology.
On each floor, the Culture of Congestion will arrange new and exhilarating human activities in unprecedented combinations. Through Fantastic Technology it will be possible to reproduce all "situations" – from the most natural to the most artificial - wherever and whenever desired.
Each City within a city will be so unique that it will naturally attract its own inhabitants.
Each Skyscraper, reflected in the roofs of an endless

层。

我们在所有路口看到天桥，和拱廊一样宽，有着坚固的栏杆。我们看到城市的小公园（我们相信比现在会多的多）被提升到和拱廊人行道一致的高度……全方面变成非常现代化的威尼斯所拥有的，一个拱廊、广场、桥梁的城市，运河作街，只是运河没有注入真正的水，而是自由流动的车流，而太阳在车子的黑色顶棚上闪耀，建筑倒映在快速滚动的车辆构成的起伏洪流上。

"从建筑学角度看，就形式、装饰和比例而言，这个想法展现了威尼斯所有的美好之处，以及更多。没有任何不协调的地方，没有任何奇怪……"

科贝特关于纽约交通问题的"解决方案"是曼哈顿主义历史上最为明目张胆的诡诈案例。实用主义被如此扭曲至纯粹的诗歌。

理论家其实并无一刻想去纾缓拥挤。他真正的野心是将之升级到这个程度——如同量子跳跃——以生成一个全新的状态，即拥挤成为神秘的正面因素。

远不能解决任何问题，他的提案是一个整理和解释一个其实令人费解的大都会的隐喻。

伴随这个隐喻，曼哈顿许多潜在的主题被证实了：在科贝特的"非常现代化的威尼斯"，每个街区变成一个拥有自己灯塔的小岛，费里斯的幻影"房屋"。曼哈顿的人口——从一个街区旅行到另一个街区——最终并确实地在2,020个这个城市自己形成的岛屿所构成的大都会环礁中定居下来。

拥挤

费里斯、科贝特和区域规划的作者们发明了一种以理性来处理根本上的非理性的方法。他们本能地知道如果想解决曼哈顿的问题那就相当于自寻死路。他们知道他们托这些问题的福而存在，也是他们的责任使这些问题，要说有什么的话，永远不能克服。他们知道曼哈顿唯一的解决方案是对它的畸形历史的外推，那就是曼哈顿是一个航班永远向前的城市。

这些建筑师的规划——合成于地区规划委员会——一定在反其目的而行之。它包含了强加于曼哈顿所释放的爆炸性物质之上的一连串的隐喻模型——它原始而又高效——替代了文学组织——在其他任何情况下都绝不可能——形成一种诗性控制的形式。1916年区划法规中的"房屋"和"村落"，费里斯的"像纪念物的建筑"，最终还有科贝特的像"非常现代化的威尼斯"一样的

不同的观点 · 31

曼哈顿，一起形成了趣事构成的却是极度严肃的矩阵，形成了一个诗性公式的词汇表，以隐喻性规划这一新学科来取代传统的目标性规划，来应对大都会中那些本质上不可量化的情况。

拥挤本身就是实现网格现实中的每个隐喻的必要条件。只有拥挤才可以生成超级住所，巨型村落，雄伟山体，还有现代化的汽车威尼斯。

合在一起，这些隐喻是拥挤文化的基础，而这才是曼哈顿建筑师们的真正活动。

文化

拥挤文化提倡由单一结构来征服每个街区。

每个建筑将转变成为一个"房屋"——一个私密的领域，被膨胀到可以容纳客人的程度，但又不至于到在所提供事物的范围内矫饰普遍性。每个"房屋"将代表一种不同的生活方式和不同的意识形态。

在每一层楼，拥挤文化将会以前所未有的组合来安排新的、令人振奋的人类活动。通过神奇技术将有可能重现所有"情况"——从最自然的到最人工的——无论何时何地，只要有此欲望。

每个城市中的城市将会如此独特，它自然地会吸引它的居民。

每座摩天大楼，倒映在无尽车流中的豪华黑色加长礼车的顶棚上，就是那个"非常现代化的威尼斯"中的一个小岛——一个 2,028 个荒岛构成的系统。

拥挤文化是 20 世纪的文化。

《谵妄的纽约》PP110-125
Rem Koolhaas
New Edition, 1994
The Monacelli Press
此段卓旻译

flow of black limousines, is an island of the "very modernized Venice" - *a system of 2,028 solitudes*. The Culture of Congestion is the culture of the 20th century.

Delirious New York, PP110-125
Rem Koolhaas
New Edition, 1994
The Monacelli Press

山水城市

History constitutes the identity of a city, while that identity of China's cities was severely dissevered through the recent decades of great urbanization, the trend of which a few struggled, but failed to buck. The cause of this rift could be attributed to the subconscious of inferiority of the whole nation stimulated by the weakness accumulated over the modern history – those belong to the ancestors have degraded us for years are naturally inferior, and anything otherwise opposite, as imported stuff, must be legitimately superior. After China's reformation, the sudden impact of visual gap between China and West greatly strengthened this sense of inferiority, thus impelled resolute rupture with the past.

At the time that history is gradually abandoned by itself, commercialism hastened by globalization rapidly hijacks the city's development. Commercial culture's genetic pursuit for profit determines that city based upon it would ultimately develop into a form of super-density, with shortsighted interest. As the phenomenon observed by Rem Koolhaas in his "generic city", cities driven by capital flow have become increasingly homogeneous. With the advance of urbanization, McDonald's and Starbucks have become an omnipresent being, taking place of traditional local snack bars, and so is the life style of the masses. This homogeneous trend on one hand helps relieve strangers from alienation to a new city, it however produces a lack of spiritual sense of urban space as well as a deep sense of loss and nostalgia for "the past world".

Shanshui City
Guo Xi, famous painter of the Song Dynasty, stated in "*Elegance of Woods and Brooks*", "There is in the Shanshui, to be walked, to be sighted, to be dwelled, to be travelled…though walking and sighting are not equally enough to dwelling and travelling in Shanshui." That is more than Ancient Chinese painting theory, it is rather their cosmopolitan vision. It is particularly worthy of taking a fresh look at this high ideal of living among Shanshui, when in this transformation process of Chinese cities' urban fabric from courtyards and alleys to skyscrapers and boulevards.

　　历史传承是城市的身份认同，最近几十年的中国城市发展体现的却是一种从上而下的身份割裂，少数有识之士的呐喊对此趋势无能为力。这种历史性的割裂产生的主要原因似乎可以归咎于中国几百年的积弱所引起的民族自卑的潜意识——那些让我们没落了几百年的老祖宗的东西自然也不是什么好东西，相反的只要是舶来的自然是好的。而在改革开放之后中西方巨大的视觉落差更是大大加强了这种自卑感，从而促使毫无留恋地和过去决裂。

　　在历史性被逐渐自我抛弃的同时，全球化进程所带来的重商主义却迅速的劫持了城市的发展。商业文化对于利润的天生追求决定了由它主导的城市发展是超密度的，是短视的。如同库哈斯所观察到的"普适性城市"的现象，资本流动所推动的城市变化日益同质化。随着城市进程的推进，随处可见的麦当劳和星巴克取代了传统的小吃店，而大众的生活方式也趋于一致。这一同质化的趋向一方面让外来人对新的城市不再陌生，但是另一方面却带来了城市空间的精神意义的缺失和对"过去的世界"的一种深切的失落感和怀旧感。

山水城市

　　宋代郭熙在其《林泉高致》中有云，"山水有可行者，有可望者，有可居者，有可游者……但可行可望不如可居可游之为得"。这不仅是中国古人的画论，更是其世界观。在中国城市肌理从小巷院落向大路高楼迅速转变的过程中，尤其值得对中国古人居于山水之间的这种理想重新加以审视。

　　"山水"，而非城市设计领域应用度更广的"景观"。在城市理论的语义学层面有其重要意义。

其一，"山水"一词为中文独有，山水画的英文翻译"Landscape painting"其实中文相对的是景观之意。而"山水"又向来和中国文人有着千丝万缕的联系，孔子曰"智者乐水，仁者乐山"，"山水"一词毫无疑问带有极强的中国烙印。其二，"景观"多为人造，"山水"相对更倾向于自然之变，其中蕴涵着中国"天人合一"和"道法自然"的传统哲理；再深一层研读，"山水"之意有避世的态度，古有陶渊明"采菊东篱下，悠然见南山"的寄情山水归隐田园，而"道法自然"更有一种无为的态度在内，当然这个"无为"不是指无所作为，而是反对过多的人为干涉，这不啻为针对目前城市环境的过度规划和开发而提供的一个全新思路。

当今西方城市的现状从很大程度来讲归根于二战之后蓬勃发展的现代主义运动。但即便是在现代主义运动鼎盛时期，全球各地信奉国际式的建筑师们对于如何体现各自地区和民族的建筑特色的努力却未中断过，因为任何想保持一点建筑文化独立性而不湮没于国际化潮流中的建筑师都有从乡土建筑和地域主义传承和借鉴的倾向。但是地域主义这个概念却是一个悖论，任何地区的固有文化不是一成不变的，也不是和其他地区的文明相对立的，相反任何文明的发展都离不开和其他文明的交融。相对于传统的地域主义对于外来文化和形式的强烈排斥，刘易斯•孟福德指出了地域主义中的相对性。在对全球化和国际风格持批判态度的同时，孟福德对地域主义关于传统的极端定义也持批判态度。当地域主义不再被看作是封闭的，而是能够和全球化趋势相交流和沟通，这个理论瞬时变的富有生命力。孟福德是第一位在对地域主义再思考后进行系统性批判的建筑理论家，他所采取的融合而非抵制或隔离的批判性地域主义态度很好的解决了传统地域主义的悖论。

回过来分析"山水城市"的理论基础，这个提法无疑是体现中国传统的特有人文气质的，归隐于山水之间历来是中国文人向往的一种理想境界，那种寄情山水的闲情逸致也是西方文化所不能体会到的。工业革命之后19世纪西方城市环境极度恶化，由此产生了一系列的以赞美和引入田园风光为理论核心的城市改造，这其中很重要的是"Picturesque（画意派）"对于城市改造的影响，深受此影响的有西方现代景观学之父奥姆斯泰德在美国所做的大都会城市开放绿地系统。在英美的城市理论体系当中，类似的田园郊区理论和实践可以说是英美城市特别是郊区发展的一个重要指导原则。相比"画意派"的田园郊区对于纯自然景观的关注，"山水城市"的内在涵义

"Shanshui", instead of "Landscape" that is widely used in the discipline of urban design, has a significant meaning in semantics. First, "Shanshui" is a word unique in Chinese, while "Landscape painting", which is used to translate Shanshui paiting, does not reflect the meaning properly. "Shanshui" has been interwoven with Chinese ancient scholars semantically. Confucius once stated, "The wise enjoy the waters(shui), the benevolent enjoy the mountains(shan)". The word of Shanshui bears a significant brand of China's ancient ideology. Secondly, "landscape" tends to be a little more artificial in modern context, while "Shanshui" is surely a pure form of nature, which reflects Chinese traditional philosophy of "harmony between man and nature" and "Tao models itself after nature". With one layer deeper, "Shanshui" bears a meaning of retiring from the real world, as reflected by Tao Yuanming the recluse' poem, "While picking asters 'neath the Eastern fence, my gaze upon the Southern mountain rests". "Tao models itself after nature" embodies an attitude of inaction, which does not imply taking no action, but rather against too much interference, which provides a fresh angle toward the excessive planning and development of urban environment.

It is fair to say that the present western cities are mainly shaped by the development of modernism after World War II. But even in its peak time, architects all over the world embracing modernism have never ceased their effort in exploration of regional and ethnological characteristics. Anyone trying to preserve a little independence out of this tide of international style would naturally turn to vernacular architecture and regionalism for aid. But the concept of regionalism itself is a paradox, for in no area the inherent culture is immutable, or necessarily opposite to that of other regions. On the contrary, development of culture can not really occur without blending with or absorption of other cultures. An absolute pure and sealed culture is an impossible concept at its very beginning. As opposed to the fact that the traditional regionalism strongly rejects foreign culture and form, Lewis Mumford points out the relativity of regionalism. When he refuses globalization and international style, Mumford holds a critical stance toward regionalism's extreme definition of tradition as well. Once regionalism is no longer treated as a closed concept, but rather communicative with globalization trend, this theory instantaneously becomes vital. Mumford is one of the early theorists that contributes systematic critical thinking toward regionalism. His view of critical regionalism that embraces integration as opposed to segregation or confliction unburdens the paradox of traditional regionalism.

Back to the analysis of theoretical basis of "Shanshui City", this concept embodies the unique humane temperament of Chinese tradition. Retiring to "Shanshui" is the highest ideal of Chinese ancient scholars, and that unfettered leisure among "Shanshui" is so subtle that can hardly be sensed by western culture. Due to the extreme deterioration of the urban environment after industrial revolution in 19th century, a series of urban renovation based upon the theories of praising and introducing country landscape into urban texture took place in different western cities. "Picturesque" style is one of the most favorite models of that time, which had deep influence on urban projects like open park system that Olmsted, the founding person of discipline of modern landscape design, did for some metropolitan area. Similar to that, Garden City theories and practices have led the development of suburb area of western cities. Compared to garden suburb of "picturesque" style focusing on the landscape of nature, the implication of "Shanshui City" bears greater significance of cultural consciousness on tradition. The temperament of "Shanshui City" is not simply reflected by some green urban space, but rather its own context.

When developing or practicing this urban theory with features of regionalism, it is unwise and even dangerous to keep the comprehension of "Shanshui City" at the populist level. Taking a look at the status of the development of Chinese cities, it is a fact that the basic framework that shaped our cities is structured with the modernism theories from the west. For the development of global cities, modernism featured with rationality is still the mainstream in spite of its many mistakes and deficiency. Encountered by this mega-complex of urban organism, it is the very rationality of modernism that is capable of integrating dwelling, working, transportation and other urban functions systematically. Without denial of modernism as the mainstream solution to the urban development, "Shanshui City" can be recognized as a major theoretical adjustment of the development of Chinese cities to correct the blunders ever inflicted. Its importance can be equaled to the post-modernists' retrospection of the lack of humanity in modernism architecture. From this point of view, the theoretical essence of "Shanshui City" is connected with that of critical regionalism.

The Way of Shanshui City
There is no difference between theories in terms of how to depict a bright future, whereas those utopian theories always turn sour when put in real practicing. Corbusier proposed the concept of fundamental factors of fun, like sunlight, air and parks in his book "*The Radiant City*", the "City of To-morrow" as he depicted is as "recessed apartment buildings in the Radiant City. Parks and schools in the middle. Elevator shafts spaced out at optimum distance (it is never necessary to walk more than 100 meters inside the buildings). Auto-ports at the foot of the shafts, linked to the roadways…In the park, one of the large swimming pools. Along the roofs, the continuous ribbon gardens with beaches for sunbathing." This conception is undoubtedly epoch-making and revolutionary. Looking back at those urban planning of that time, who is not to be shocked by taking this sight: rows of modern housing buildings, separated by large green parks, what a modern way of living. Then we all know what the modernism cities developed into. Rigid zoning laws led to unbalanced urban development, forming obsolete urban space as well as that of overloaded usage. At the meantime, the development of transportation brings convenience at

绝不应该局限于此，而是有着更丰富的中国传统文化意味。"山水城市"的城市气质不是简单的体现在城市的绿地景观空间上，而是应该体现中国城市自身的特征和文脉的。

同时要注意的是，当发展"山水城市"这一具有地域特色的城市发展理论时，如果对它的理解停留在民粹层面则是相当不智而且危险的。纵观当今中国城市的发展现状，其发展理论和建设框架基本来源于西方的注重功能的现代主义城市理论。对于当代全球的城市发展而言，尽管有许多不足和失误，基于理性的现代主义趋势仍是主流，原因很简单，因为对于城市这么一个超级复杂的有机体，只有现代主义的理性才有可能系统性的安排城市的居住工作交通娱乐等各项功能。在确定现代主义是大的城市发展方向的基础上，"山水城市"的理论可以是对当代中国城市发展失误而产生的一个重大理论调整，其重要性犹如后现代主义对于现代主义建筑人性化缺失的反省。从这点来看，"山水城市"的理论本质应该是带有批判性的地域主义。

山水城市的解决之道

任何理论在对美好愿景的描述上毫无二致，但是乌托邦倾向的理论往往在实际操作过程中出现偏差。在《光辉城市》一书中柯布西耶提出了阳光、空气、绿地等"基本欢乐"要素的观念，他所描绘的"明日之城"是"退层的建筑，中间是公园和学校。电梯分布在一个合理的距离之间（在建筑内不需要走超过一百米的距离）。车库就在电梯附近，和道路直接相连……公园内有游泳池。屋顶花园连绵不绝，人们在上面可以进行日光浴。"这样的城市构想无疑是划时代的和革命性的，现在回过头来看看这些城市规划，不禁也为它的规模和气势所震慑。那一幢幢整齐划一的现代主义建筑，间以大面积的公园绿地，这是多么令人心驰神往的城市啊！但是现代主义城市的实际发展过程却不尽如人意。僵硬的城市功能分区带来不均衡的发展而同时形成超负荷使用和荒废的城市空间，交通运输的发展在带来便利的同时又割裂着城市的脉络。

所以当我们讨论"山水城市"，在肯定其理论的重要性之后，我们更应关注是这个理论的如何贯彻。

自然之道

"山水城市"的营造须发于自然，这里的"自然"非指自然景观，而应是自然规律，是中国传统的哲学观，是"自然之道"。在自然之道的原则下，城市中的人造物和自然物是应该最大限度地媾和在一起的，而成为一种此即是彼、彼即是此的状态。在这种中国传统哲学观的引领下，城市和乡村这两者也是模糊在一起的，"山水城市"的某种状态可能就应该是将城市消融在乡村当中，这种反城市化的倾向恰恰应该是在当下中国城市化背景下对城市认识的一种与时俱进的态度。

但是，无山无水之城而要硬造山水是违背自然规律的，也是和当地的文化特征相冲突的。中国山水重于意而轻于形，唐代王维《山水论》有云：凡画山水，意在笔先。同样，以意境表达山水不失为营造"山水城市"的一个重要手段。"山水城市"的意境是自然的、优雅的。务必要避免挂"山水城市"之名，或是企图复制自然，或是大造华而不实的人造景观，中国的城市已经有太多这方面的教训了。

现代都市里高楼林立，要造山水难矣，倒是不妨从便宜之处入手，比如见缝插针多造林木。清代钱杜所著《松壶画忆》即有"山水以树始"的论断。所以，并非建筑才是改变城市的手段，植树恰恰是"山水城市"的设计当中最为重要的一环。

地域性

即便同为山水，北方山水和南方山水也大异其趣。传统中国文人画中的山水普遍表现的是江南山水——山色空蒙而云水缭绕，而北方山水则是斧劈刀削而大开大阖。神州大地，天南海北，各有其地域特征，是否都要以江南山水为模本进行复制呢？答案当然是否定的，"山水城市"的精髓不在于表面的"山水"，而在于其后的地理和文化的地域性特点。如果是江南的城市，那当然就可以充分吸收苏州园林之美；而如果是黄土高原的陕北，要移植江南的山水景色是不可能也不可为之事。所以，"山水城市"是一个宏观的城市理念：胡同和四合院是皇城根儿的"山水"；成排的窑洞就是黄土高坡的"山水"。

同时，"山水城市"的批判性地域主义特点决定了它不应是固步自封于传统当中的，因为即使传统本身也是在与时俱进的。继承历史不代表忽视当下，历史积淀所形成的文化特色是要有机的融入到新的城市发展当中去才有其生命力的。如果只是切割出历史的某一部分不加创造的重新组装和拼接到现有的城市文脉当中，比如当今一些城市新建的画虎类犬的仿古街道，那不仅是低级庸俗的，也是对城市认知的巨大威胁。

the cost of rupture of urban texture.
To make the theory more practical, it is important to carry out the discussion of implementation of "Shanshui City" as its significance is affirmative.

Tao of nature
The making of "Shanshui City" must be originated from nature, which, instead of natural landscape, refers to natural laws. That is Chinese traditional philosophy, a "Tao of nature". Guided by this principle, urban artefacts shall be blended in the natural environment to the maximum extent, in order to reach a state without distinction. Guided by this principle, the boundary between urban and rural is blurred. Certain state of "Shanshui City" could be a city disintegrated into the rural. This anti-urbanization inclination is most likely a responsive and progressive attitude in the background of China's great urbanization.
It is against the natural law as well as local geographic and cultural identity to create "shan" and "shui" where there is no such conditions. Chinese "shanshui" puts more weight on the conception than form. Wang Wei, the famous Tang Dynasty painter, said in *Shanshui Talk*, "when painting, conception comes before the brush touches paper." Similarly, conceptualization of "Shanshui" in unban context could serve an important role of making "Shanshui City". This conceptualization shall be natural and elegant, and must avoid any attempt of emulation of nature, or creation of artificial nature, by which Chinese cities have been given enough lessons. However, it is difficult to create "Shanshui" in a contemporary city that has been filled with high-rise buildings. A feasible way is to start from the minor yet easy part, for instance, planting trees wherever possible. Qian Du, the Qing Dynasty painter, had this conclusion of *Shanshui starts from trees* in his book "memoir of pine painting". After all, architecture is not necessarily the only way of transforming a city, yet planting could be the most important link in the design of "Shanshui City".

Regionality
Under the same term of "Shanshui", the northern one is quite different from the southern. Chinese traditional scholar paintings generally represent the southern "Shanshui" – misty mountains amidst clouds and water, while the northern "Shanshui" features an image of much larger scale with ax-chopping texture. Different territory has varied regional characteristics. Is it imperative to model the elegant southern "Shanshui"? The answer is no for sure. The essence of "Shanshui City" does not rely on the superficial, but rather geographic and cultural. A southern city surely can absorb the aesthetics of Suzhou Gardens, while for a city in the north like on the loess plateau, it is impossible and unnecessary to transplant the misty southern "Shanshui". "Shanshui City" is a macro urban concept: Hutong and courtyard are the "Shanshui" of the forbidden city, and rows of cave dwelling are the "Shanshui" of the loess plateau.
Meantime, bearing the feature of critical regionalism

determines that "Shanshui City" is not limited to its tradition, for tradition itself evolves. Historic heritage does not imply the neglect of the present, the cultural characteristics need to be integrated organically into the new urban development to extend its vitality. Any parts, easily dissected from history, assembled into urban context without creation or sophistication, like those antiqued streets, are not only distasteful, but a great threat to the mass perception of cities.

Urban craftsmen

An important factor in making "Shanshui City" is who is to lead. Engineers and planners are the builders of the modern cities. But that is only a fact for no more than a hundred years, even in the western world. They create events and change the world through structure, and cherish the faith – break through the current restrictions and conceptualize a common set of principles to lead the development of cities. This kind of principles always override natural laws and seldom bend itself before historical or geographic environment, which has been proven to be the source of dullness. Levi-Strauss proposes a new role - "bricoleur" – as a parallel one to engineer, in his study of mythology of primitive tribes. "Bricoleur" is regarded as being able to create with whatever at hands, and with far more freedom. If we know a little about Suzhou Garden, the creators are the "bricoleurs" of Chinese version – an alliance of traditional craftsmen and scholars. Although this alliance has long gone, the creation of "Shanshui City" can only be achieved through surrounding, through experience that bridges artificial and natural, as opposed to overriding the nature. It is imperative and decisive to restructure the role of craftsmen in order for the feasibility of "Shanshui City".

Urban scale

One objective observation is that large scale cities – like Beijing and Shanghai, are already in line with those of west, as well as the homogenization of city image. It is extremely difficult to infuse any "Shanshui" into the urban fabric of these westernized cities. Compared to these large scale cities, some small-to-medium scale cities are still at the early stage of urbanization and have better preservation of regional culture. They shall be the primary focus of the implementation of "Shanshui City". The issue of development needs to be reconsidered in smaller cities. Without denial of necessity of development of internal structure, whether and how to develop the external features so as to maximize the benefit of smaller cities need to be remeasured.

Limited to the scale, it is easier to integrate the concept of "Shanshui City" into smaller cities from the angle of planning. By achieving a properly planned "Shanshui City", smaller cities would not only build the identity, but substantially improve the competitiveness and thus attract intellectuals so as to alleviate the population pressure in larger cities. From a macroscopic perspective, it is of strategic significance to promote "Shanshui City" in smaller cities in the background of China's urbanization.

城市工匠

在营造"山水城市"的时候，一个重要的因素是谁来主导。工程师和规划师是现代主义城市的建造者，但即使在西方世界，这也不过就是一百多年的历史。他们通过结构来创造事件和改变世界，他们怀抱着这样的信念——必须突破现有的限制从而构思出一种通用的法则来引导城市的发展。这类通用的准则总是试图凌驾于自然至上，极少因历史或地理环境而变通，这在中国近三十年的城市化进程中已被证明是一切无趣的源泉。列维-斯特拉斯在对原始人的神话学研究当中，提出了和工程师平行的"修补匠"的概念，认为"修补匠"总是就手边现有之物来进行的，更为随心所欲。如果我们对于苏州园林有所了解的话，其创造者就是中国的"修补匠"——传统的工匠和文人。尽管这一合作体系在当代早已崩溃，但是"山水城市"是需要根据周遭之物，以经验来联通人工和自然之间，而非凌驾于自然之上。所以重建工匠这一角色对于"山水城市"的可行性具有决定意义。

城市尺度

现在的客观事实是大型城市——如北京上海等，其开发程度已可比肩西方，城市意象也趋向于和西方城市同质化，要在现有的城市肌理和脉络上再添加中国特色的"山水"意境绝非易事。相比大城市的积重难返，倒是中小城市正处在城市化发展初期，而且相对大城市而言，地域文化的保留相对要好一些，所以中小城市应该是实施"山水城市"的重点。中小城市的发展应该重新认识"发展"这个问题。当然我们绝不否认内在结构的发展必要性，但是外部城市特征的是否发展与如何发展才符合城市的最大利益却是需要重新衡量一下的。

由于城市尺度有限，中小城市更易于从整体规划的角度把"山水城市"的理念融合进去。如果中小城市能够建设得当的"山水城市"，不仅能明确城市本身的定位，也将大幅度提高城市的竞争力，吸引人才，从而在另一方面减轻大城市的人口压力。所以从宏观角度来讲，在中小城市建设"山水城市"对于整个中国城市化的进程是有战略意义的。

结语

亨廷顿在《文明的冲突》一书中指出冷战之

后国与国之间的对立有着向不同文明之间的对立所转变的趋势，在此本文无意讨论亨廷顿的文明冲突论。但是，文明作为一种存在，中华文明是否能在现代社会不被西方文明淹没，这不仅取决于延续的年代或是辐射影响力，还看这种文明是否真正能在协调、融合中坚守。"山水城市"理论在中国的城市规划和设计领域触及了我们或许不敢触及的东西——中华文明的独立性，但作为专业人士，我知道这点是我们所绕不过去的。

卓旻

Conclusion
Huntington points out in his book "*Clash of Civilizations*" that confrontation between nations trends toward between civilizations after the cold war. This essay has no intention getting involved with the discussion of the clash of civilizations, though it needs to be aware whether civilization of China would be devoured by that of west relies upon not only the time it lasts or the influence it once had, but whether this civilization could hold its own essence among cultural integration. "Shanshui City" theory touches something that have been neglected deliberately in the discipline of urban planning and design in China for years – the independence of Chinese Civilization, yet we know any attempt to bypass this topic is in vein.

Zhuo Min

时代的案例

CASE STUDIES OF AGES

吉斯工人住宅中庭

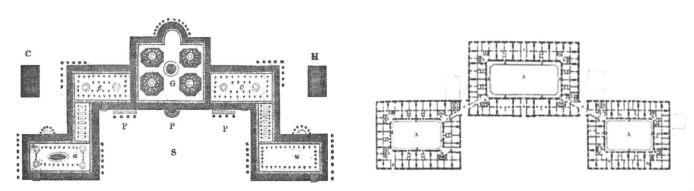

法郎斯泰尔平面

吉斯工人住宅平面

吉斯工人住宅与工厂

40 · 时代的案例

理想公社

19世纪的欧洲,工业革命将大量的农民转变成城市中的产业工人。人口的大量涌入使得劳动者在城市中的生存条件极其恶劣。在踏入工业时代初期的欧洲城市,城市工人集合居住的理想公社这一概念诞生的意义远远超过一种建筑类型的发明,而是代表着一种社会模式的进步。

空想社会主义者傅立叶让建筑师为他的理想社会设计了一个集工作劳动和生活居住于一体的城堡式理想公社——法朗斯泰尔[Phalanstere]。在此概念的影响下,从1856年到1859年,法国人高登[Godin]开始开发吉斯工人住宅[Familistère de Guise]的计划。他的意图是提高职工居住标准,并且为他们在一个综合体内提供生产、贸易、教育和娱乐等功能,形成一种可能的自给自足的社区体系。

吉斯工人住宅中庭舞会活动

吉斯工人住宅中庭幼儿活动

时代的案例 · 41

马赛公寓

"要么革命,要么建筑!"柯布西耶深刻洞察到城市当中为劳工阶层解决居住问题的主要性。居住问题是其所鼓吹的现代主义城市规划的核心问题。马赛公寓[Unité d'habitation ait Marseilles]代表了柯布西耶对于住宅和公共居住问题研究的高潮点,更是沟通其城市与建筑观点的桥梁。

马赛公寓对于人的尺度、人与居住的关系、私密性与公共性之间的关系、城市与建筑的关系等方面作出了巨大的阐述。他所提出的建筑五要素在此发挥得淋漓尽致,其内部就像一个微型城市,以让居民尽可能地接触自然和社会,并增进居民间的相互交往。

现代主义风格不为历史所羁绊,对于集合住宅而言似乎是天生的绝配。这点对于中国城市居住社区的风格讨论有重大意义。

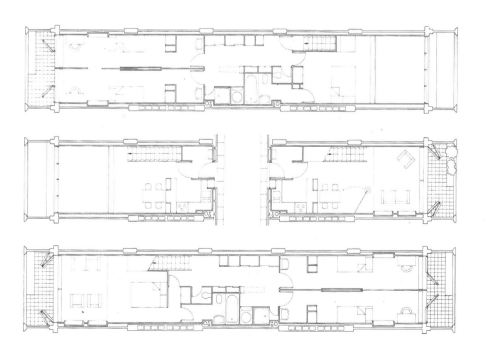

马赛公寓单元平面

老区新楼

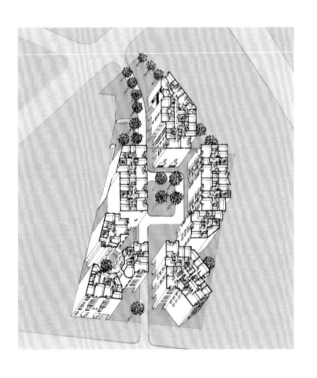

巴黎十三区是巴黎一个较为特殊的城区,这里种族多元但居民收入较低,社区环境也相对较差。在20世纪60年代现代主义规划盛行之时,这里也曾夷平大片土地重建居民区,所以巴黎其他区难得一见的高楼加起来恐怕不如十三区。

20世纪70年代,克里斯蒂安·德·波赞巴克[Christian de Portzamparc]在巴黎十三区设计的"高层形式"[Haute-Formes]公寓项目(209套公寓)展示了当代建筑如何放弃暴力拆除的手段而优雅地改变老城的面貌。

该项目摒弃了现代主义的高大的纪念物形式,而是以典型的宜人尺度的巴黎住宅街区的形式介入这个形状不规则、开口只有四十多米的特殊地块。传统巴黎住区通常是外围的建筑围合中央开放的内庭。波赞巴克设计了八个独立的塔楼围合成一块公共空间。同时建筑之间运用的并不是重复的元素,而是通过设计不同尺寸和形状的体量和空隙、不同类型和尺寸的阳台和窗户来表达不同建筑之间的区别与联系。近景和远景在每处空间和公寓中共存,通过外立面的节奏感、空地上的绿树、以及面向城市的开放,每个公寓都能感受到这里平静和充满光感的建筑景观。

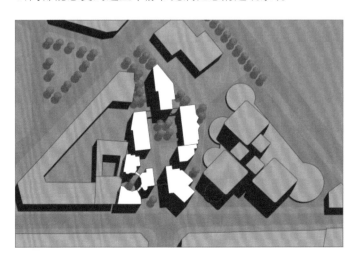

时代的案例 · 45

利塔街住宅群的不同内庭和立面

克里尔设计的利塔街北块中心立面

46 · 时代的案例

类型研究

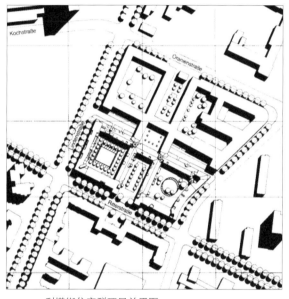

利塔街住宅群项目总平面

在强调功能单纯分区为特征的现代主义城市规划的背景下，类型学被引入建筑与城市设计的讨论之中。城市的基本类型元素被抽象出来，以便于重新组织形成可以被清晰阅读的文本。在对欧洲传统城市的纪念性和艺术性的研究中，罗伯·克里尔[Rob Krier]对在现代化过程中逐渐丧失的传统城市空间的两个基本要素——广场和街道——予以强调，他认为了解所有能想到的城市空间的类型和公共空间的各种可能的立面设计是城市设计必要的先决条件。

20世纪80年代，在柏林国际建筑展机构[IBA]的支持下，克里尔主持了利塔街[Ritterstrasse]住宅群项目，为该项目规划了连续的带有内庭的住宅区块，体现了住宅向城市街坊空间结合的倾向。沿街的新住宅楼的设计集结了近二十个建筑设计团队进行分段设计。新建筑考虑了大量的平面类型以及多样化的居住要求和生活风格，一栋建筑中将允许不同的房型和各种群体的存在。同时区域内将提供各种公众公园、街区公园、广场和林荫道，还有半私密的庭院和组团花园，产生一种相互影响的积极空间和绿化，绿化邻近住宅嵌入现有的城市结构而不是去破坏它。

克里尔对于住宅转角类型的研究

时代的案例 · 47

公共组屋

组屋这一居住社区形式是亚洲国家通过公共政策引领塑造城市面貌的典型案例。新加坡组屋始于20世纪60年代，当时新加坡刚从英国殖民统治下脱离，整个社会发展比较落后，政府财力有限，民众住房条件比较差。"居者有其屋"是组屋开发的首要且或许是唯一的目标，而其他任何塑造城市观感的因素包括传统、风貌、尺度、交流等在内则基本可以忽略不计。

组屋社区一般来说高层林立，居住单元从早期的两房到后期五房甚至六房，各种平面布置都有，外墙常用彩色涂料。组屋社区遵从现代主义规划的功能布局，一般交通便利，公共配套设施较完善。到21世纪的城中村模式，最终形成功能完备的都市聚落。

组屋在住区模式、交通系统以及基础设施的布置方面都对中国当代大量的公寓小区有着巨大的影响。

生长社区

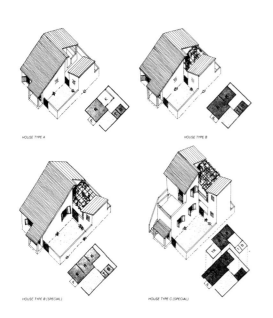

查尔斯·柯里亚[Charles Correa]试图用生长社区这一概念去证明如何利用低层的建筑居住模式去解决高密度的居住问题（每公顷500人，同时包含开放空间与学校）。贝拉普尔[Belapur]项目源自于社区空间的阶级问题以及大量移民所带来的居住问题。考虑到这个问题，柯里亚利用大约8米×8米的小型共享庭院来组织不同的居住空间，每7间房都会带有一个小型的庭院。

每户家庭都拥有其公共空间和露天庭院，这不仅可以增加公共活动的机会，而且在居住过程中能够不断的扩张空间，形成自我生长的特征。更重要的是，每户都拥有私密空间与公共庭院，是一个完整独立的居住单元，互相之间不会发生干扰，这使得这些房子都拥有了独立生长与自发建筑的潜能。

整个社区包含了不同收入等级的住户，然而为了保持公平原则，不同等级住户的面积变化并不是很大。整个社区的形式和平面都较为简单，以便住房可以在传统工匠的改造下，自我建造与自我延伸，从而也产生了雇佣关系以促进城市经济的发展。

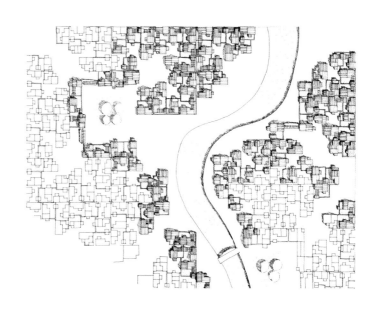

时代的案例 · 51

两块 场地

TWO PIECES OF LAND

场地一

之江
地块

课程简介

课程目的：本课题关注当今城市化进程背景中的城市集合居住的乌托邦形态。希望学生对于居住类型、城市形态、发展模式等概念有深入了解。

本实验课程试图打破建筑教学不同课程之间的障碍，从地理、历史、伦理、出行、环保、技术、政策等各个不同角度来帮助学生理解居住的意义、了解城市形态和生活模式之间的相互刺激的关系，从而培养学生的全面思考问题的能力。

设计背景：十年间，全国县以上的规划新城数量不断上升，已达几千之数。曾经的万顷良田已不复耕作，一幢幢的公寓楼正在其上拔地而起，曾经和这片土地向连接的记忆和情感却已荡然无存。

本设计课程的场地即是中国规划新城的缩影。这一类新城较之西方卫星城市大发展时期的卧城犹过之而无不及，伪西式的住宅立面不能掩盖其实质的虚无。在这一片文脉已被连根拔除的巨大空地上如何建立新来者的归属感？

课程时间：2012年春季

场地介绍

之江度假区：
十年之前杭州郊区的乡村地区，现在是杭州西南面迅速发展的城市副中心。区内西侧是转塘镇，中国美术学院象山校区所在。区内东侧是钱塘江南端，北侧是珊瑚沙水库。联接此区域和滨江区域的之江大桥在建。

Zhijiang District
The rural area located at suburban Hangzhou 10 years ago, now is planned as one of municipal sub-centers. The west part of the district is Zhuantang Town where CAA's Xiangshan Campus is located. Qiangtang river runs along this district at its east side, and a reservoir is located at north. A bridge is under construction to make this district more closely connected with the main city.

地块面积：160,000 m²   Site Area: 160,000 m²

主要交通路网

两块场地 之江地块 · 59

2000

2007

2010

地形重构
GEOGRAPHY RECONSTRUCTED

全程中方组员：顾晓丹、王祺雯、翁婷玉、周振宇、钟祺祥
中期法方组员：Arthur Bel, Yuyu Yang, Keny Ega-Bourgeois

一、将场地中原有道路的肌理保留并延续，以菱形格子作为场地的控制规划模数。
二、置入有机分布的道路、田地、水体、山体，形成新的地形系统。
三、以圆的形态统合不同的元素，形成规则的形式系统。
四、叠加圆的形式系统和有机的地形系统，完成地形重构。

田地的置入：
三种形式：最基本的产业式农田、景观呈现的特色农田、居民活动的交流型场地场。
水体的置入：
两种形式：一种呈圆形或圆环，作为景致存在；另一种呈带状环绕田地，具有灌溉农田的作用。
山体的置入：
两种形式：借鉴了福建土楼形态的梯田式建筑；隐喻山峰的错层高楼。

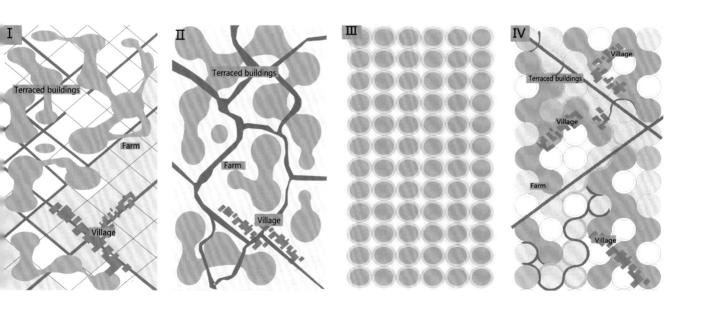

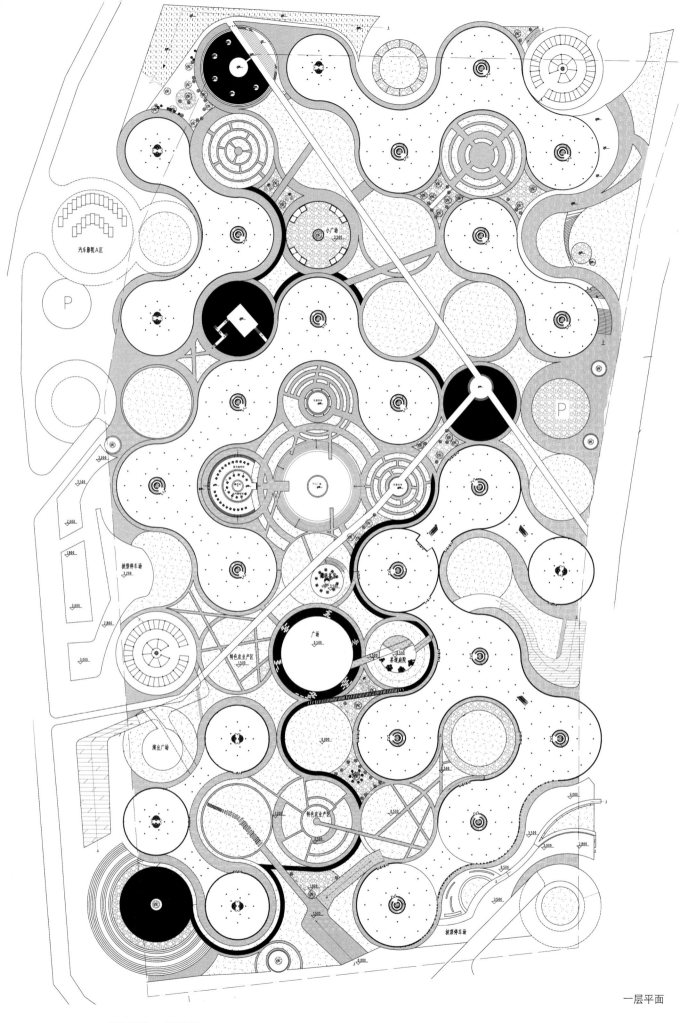

一层平面

66 · 两块场地 之江地块

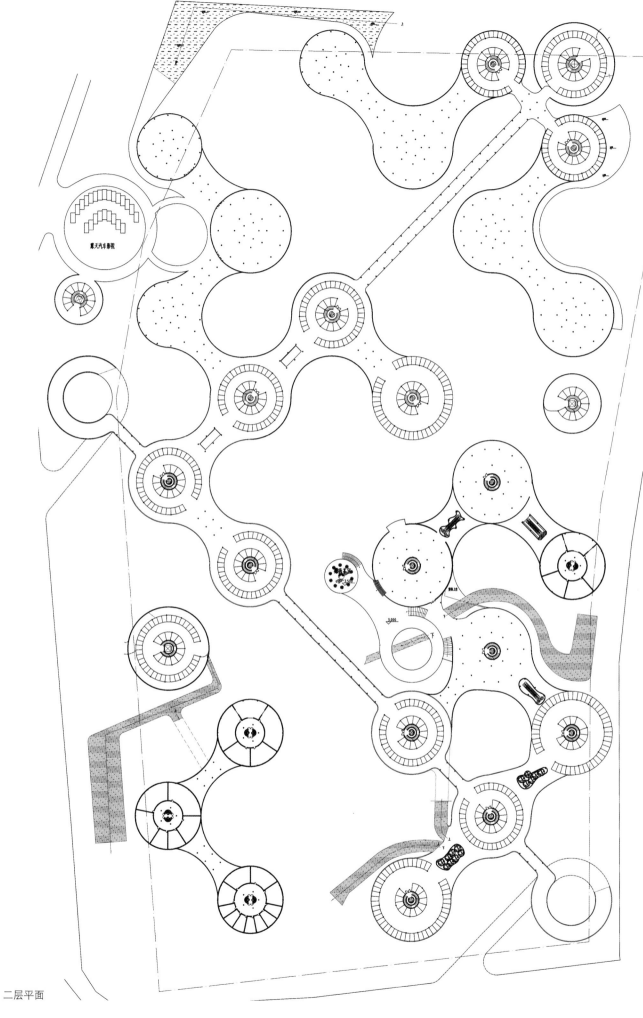

二层平面

两块场地 之江地块 · 67

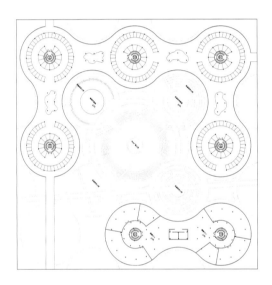

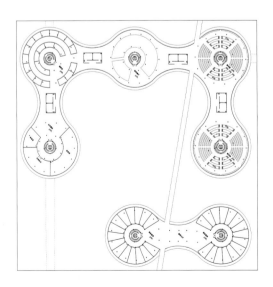

68 · 两块场地 之江地块

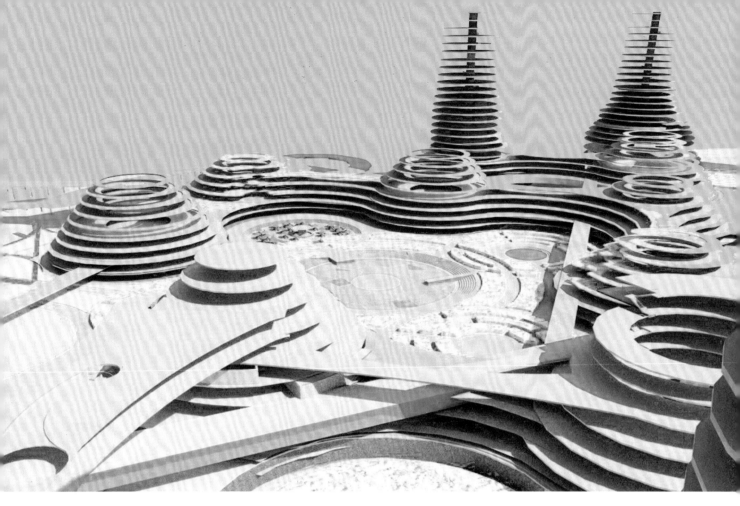

两块场地 之江地块

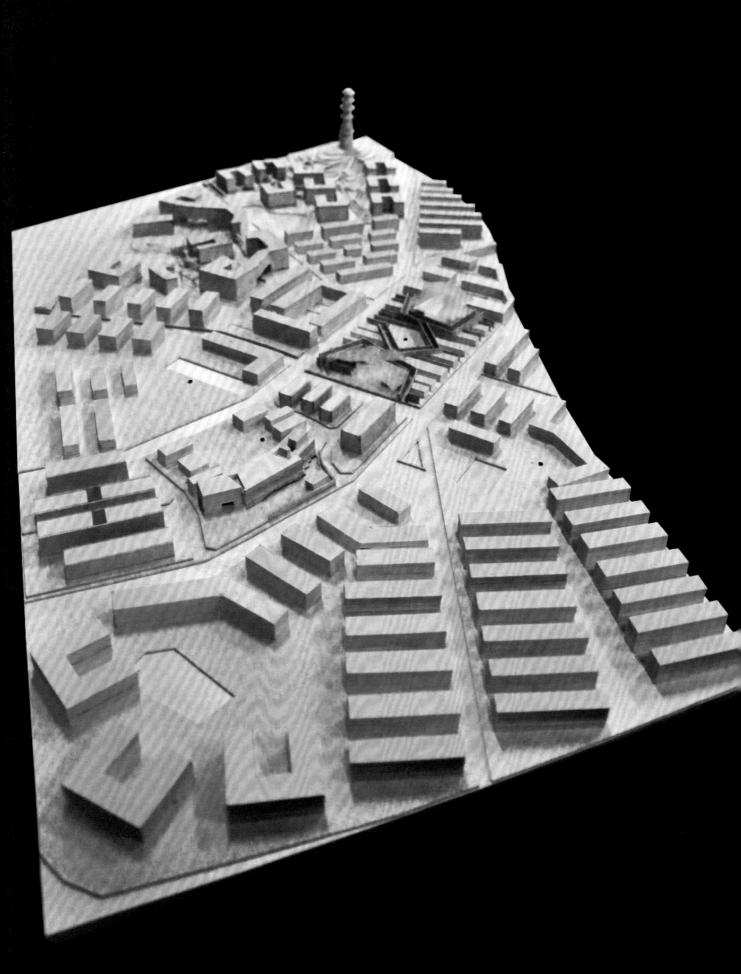

市井重塑
Remaking "Shijing"

全程中方组员：陈念海、陈江东、赵倩文、王晶焱
中期法方组员：Caroline Ebert, Marion Le Coq, Audrey Morinet

杭州为南宋故都，自古市井生活丰富多彩。旧时的十座城门皆有相应的市井生活勾连，形成各具特色的城市场景。当代城市生活的最大问题是城市各方面的均质化，不断重复拷贝的城市场景和日常生活令人沮丧。城市设计要同均质化的倾向抗争，而历史语言是一条可行的路径。
传统的中国市井中的建筑语言被引入设计作为城市肌理的密度集聚点，从而打破城市的均质化发展。

市：商业生活聚集之处，居住社区的人流中心。
戏：文化生活聚集之处，戏台控制周围的公共空间。
园：以造园手法进行公共空间与私密空间的转换处理。
塔：景观视线聚焦之点，亦为景观流线的汇聚点。

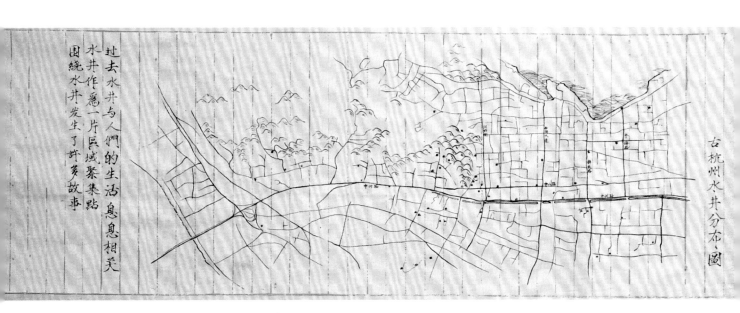

从武林门至湖墅一带，历来是杭嘉湖淡水鱼聚散地，因此杭谚有"百官门外鱼担儿"。

Ponds/ fishing

往灵隐天竺进香之人，必由钱塘门出入，故有"钱塘门外香篮儿"。

Temple/ pilgrimage

涌金门历来是杭州城里到西湖游览的通道，在古代就有游船码头，西湖游船多在此聚散，因而有"涌金门外划船儿"。

Lake/ boating

古时候市民需用柴炭多从此门运入，故有"清波门外柴担儿"。

Wood/ firewood

凤山门是连接江干一带和游览西湖的交通要道，风景优美，成为游人骑马踏青之处，因此有"正阳门外跑马儿"。

Horsetrack/ outing

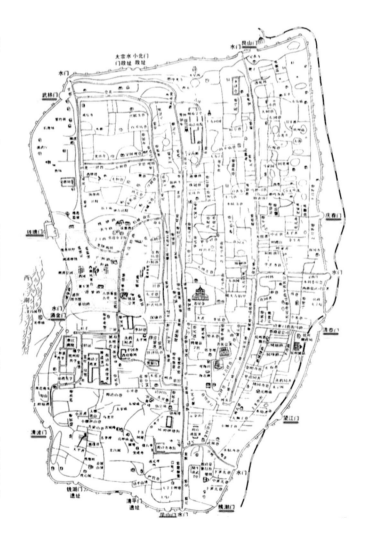

这里个体丝织户与机纺作坊遍布，机纾之声，比户相闻，因此杭州民谣有"坝子门外丝篮儿"。

Workshop/ silkspinning

门外为郊区农民的菜地，菜农运菜进城，担粪出城，均由此门出入，故有"太平门外粪担儿"。

Farmland/ manure

清泰门外沿江一带直至江水入海处，是古代煮海盐之处，沿江多盐，因而有"螺蛳门外盐担儿"。

River/ seasalt

这里乡民以种菜为业，然后挑到城里来卖，因而有"草桥门外菜担儿"。

Farmland/ vegetable

杭州城内的绍兴老酒都由候潮门入城，因此，杭谚有"候潮门外酒坛儿"。

Winery/ wine

场所/事件
由与自然环境契合的历史肌理所勾勒

场所/事件
通过对当代均质格网的变异以重塑

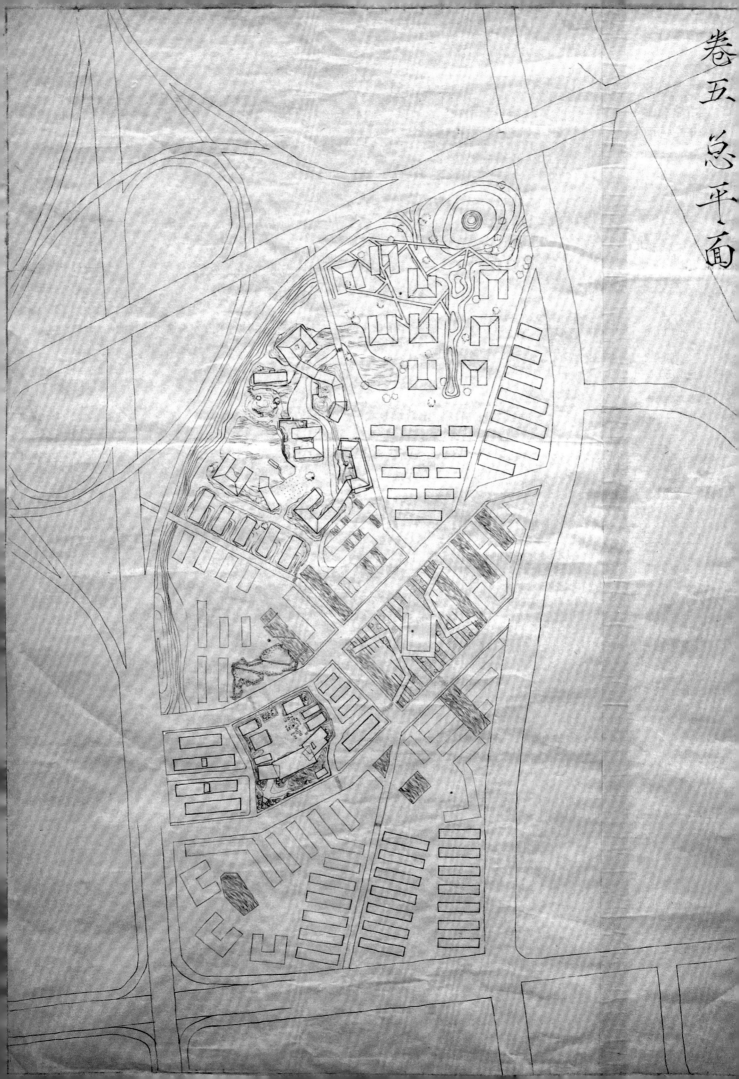

卷五 总平面

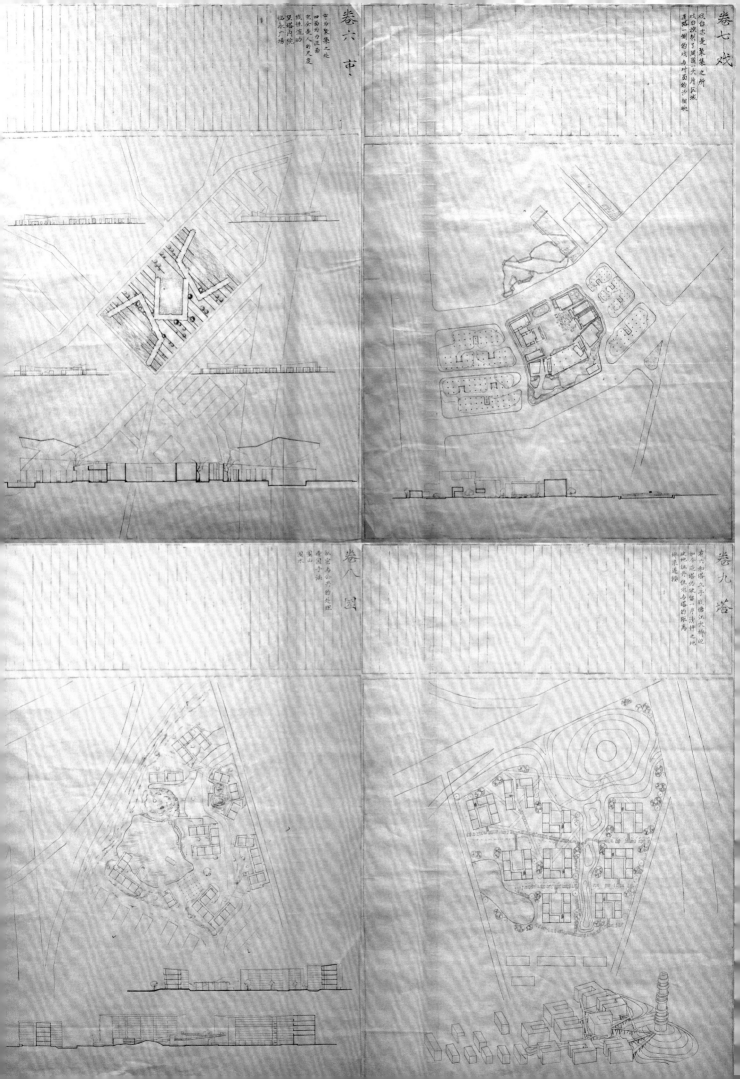

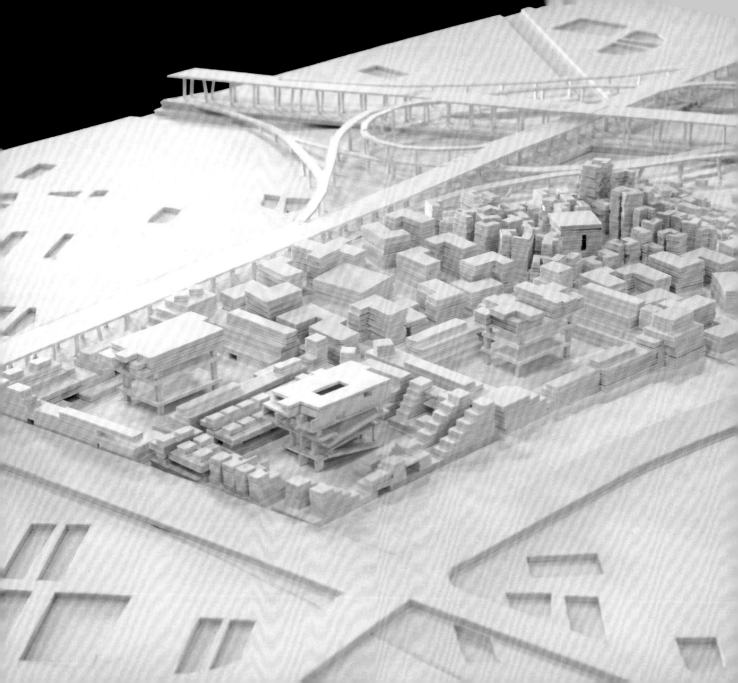

里坊相望
BLOCK ON WATCHING

全程中方组员：张艺露、林杰、王莹、陈烨
中期法方组员：Louis Bordenave, Yuan Shen, Marc Johnson

里坊制度是一种中国古代的城市规划制度，以唐代长安为盛。全城被分割为若干以高墙环绕封闭的"里"作为居住区，商业与手工业则限制在一些定时开闭的"市"中。
当代中国大量被开发的封闭居住小区何尝不是一种里坊？但是当代生活毕竟与古代大有不同，当代的居住里坊如何能加强社区居民的人际交流呢？

居住：将居住功能置于社区四周，只提供纯粹私密的居住环境，形成社区对外的封闭。
社交：放大社区中心的功能和体量以鼓励居民之间的交流，将其置于社区的中央，形成社区对内的开放。

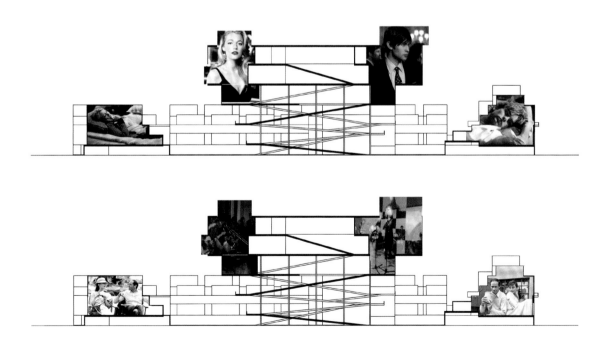

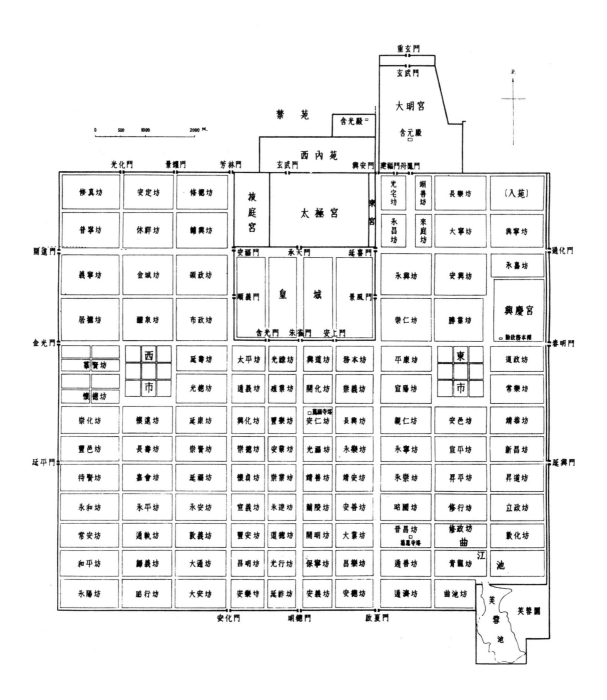

唐长安的里坊制平面

两块场地 之江地块 · 79

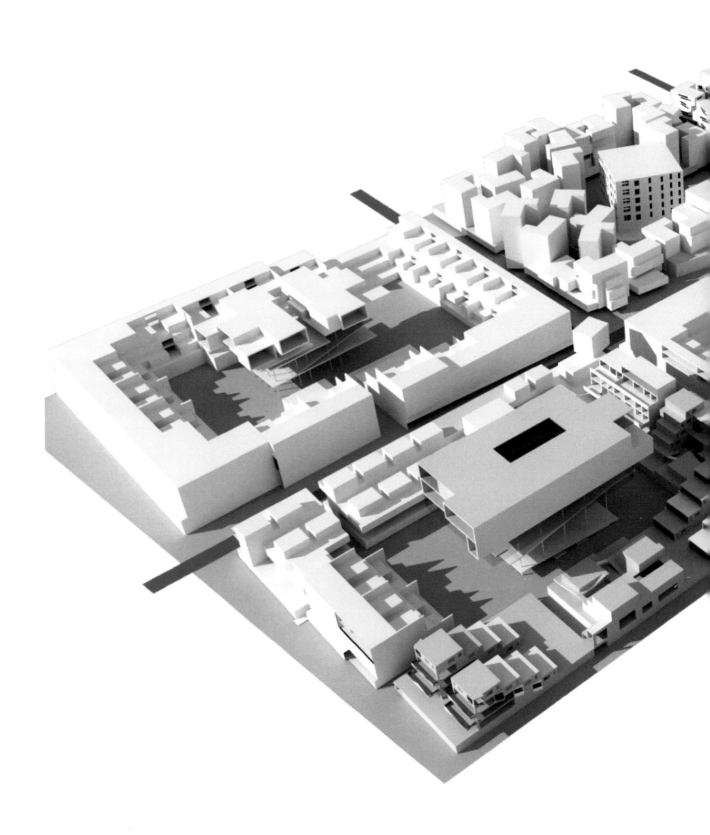

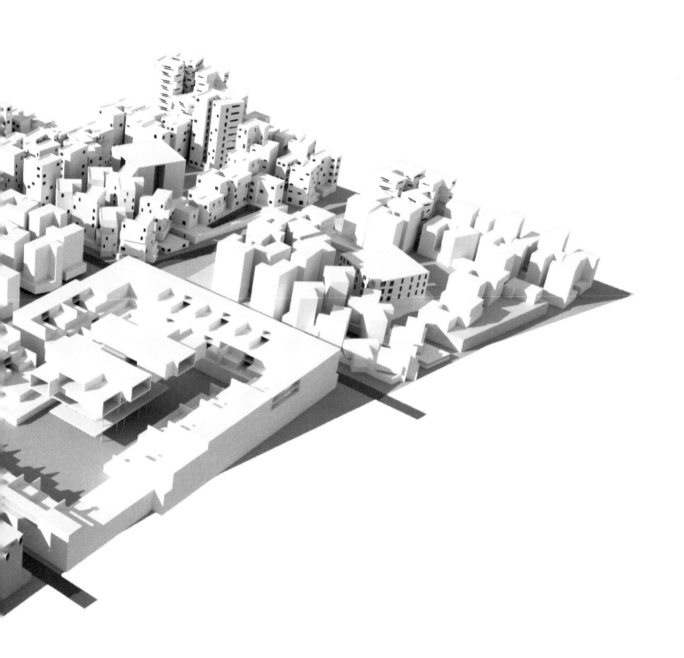

两块场地 之江地块

交通导向
TRANSIT ORIENTED

全程中方组员：隋明玉、苏明琦、童超超、赵雄斌
中期法方组员：Sophie Boehly, Nicolas Bouisson, Julien Pilon

柯布西耶认为交通是现代主义城市的核心问题。尽管现代主义对于城市分区和功能组织的机械表达受到诟病，但是交通出行毫无疑问是当代中国人口稠密的城区的最核心问题。与柯布西耶将汽车交通看作城市的发展方向不同，中国城市的人口密度不可能承受私人汽车这一单一的出行方式。客观情况是随着中国汽车产业的发展，中国的城市交通已面临崩溃的境地。

多层次的公交系统是中国城市的发展导向。城市居住社区的发展，从密度控制到单元组团，都可以依附于一种经过精心设计的交通模式进行控制。

具体交通模式组织：
交通方式的选择：城轨、公交、航运、汽车、自行车、步行，等等。
换乘方式的组织：城轨站点、大型汽车库、公共自行车站，等等。
出行范围的控制：将住宅、商业、办公、公园等设置在步行可达的公交站点的范围内。
步行网络的通达：构建适宜步行的街道网络，将社区单元有机连接起来。

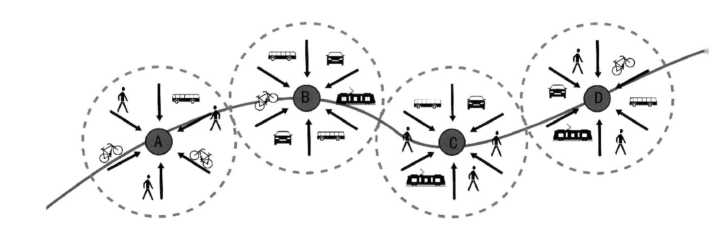

中心发散

环型蛛网

传统里弄

疏密网格

规整网格

交通模式研究

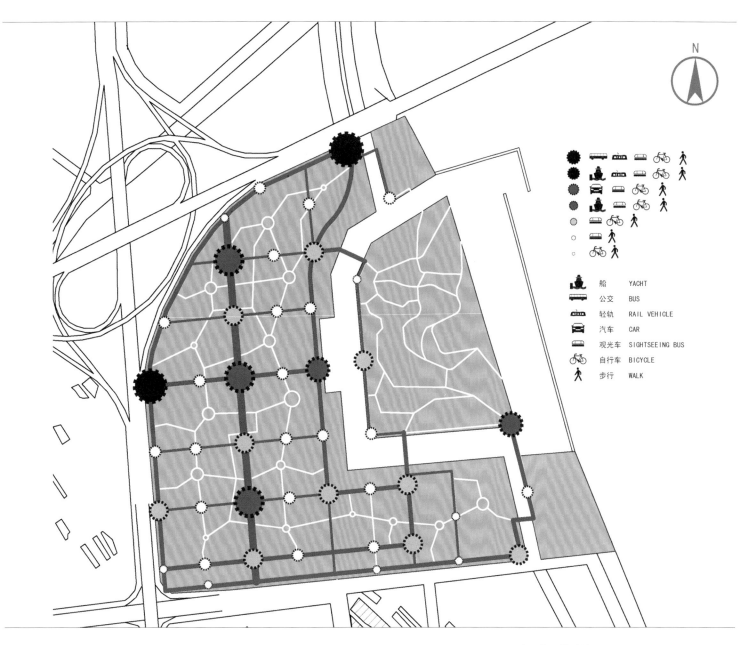

交通换乘模式总图

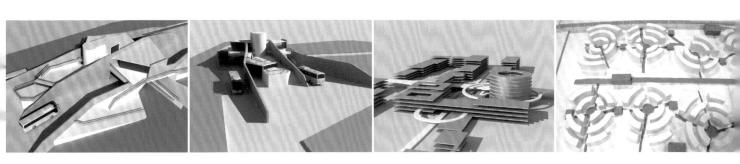

两块场地 之江地块 · 85

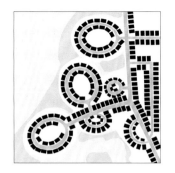

院落围合

树枝分杈

中心分级

网状并行

社区组团研究

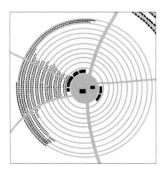

中心汇聚

社区组织模式总图

两块场地 之江地块 · 87

自然资源
Utilization of Nature

全程中方组员：王思达、魏浩然、刘思源、俞琴、周舟
中期法方组员：Adrien Lazennec, Helen Prada, Guillaume Le Moal

人居环境离不开自然的融洽，这是基本的人性需求。自然资源既提供人居生活的动能、也提供愉悦身心的美景。尤其在那些本来就具有自然资源优势的地块，如不加以利用自然资源并形成设计的导向，那是暴殄天物。

这块场地紧靠大江，而杭州夏日炎热，降温隔热是杭州地区住宅的热工设计的最主要考虑因素，所以江风的引导对于居住组团的控制和建筑形制的选择有很大影响。

这块场地原先水塘密布，从景观角度到热工角度都可善加利用。场地西南角的大水塘保留下来作为居住社区的主要水景，其他水塘通过新开挖水渠相连。居住单元靠近水塘半围合布置，将水景纳入单元形成公共活动空间。另外靠近建筑的水体也成为水源热泵的潜在散热体。

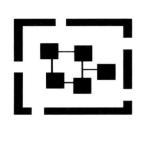
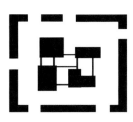

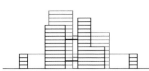

总平面示意图

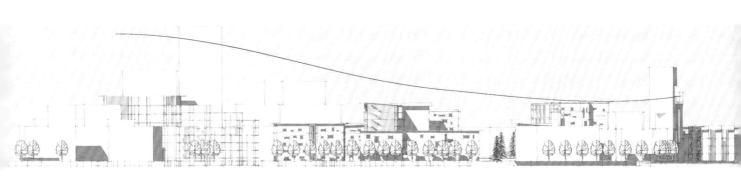

江风的引导:东西立面

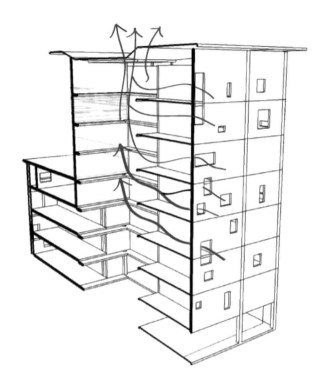

江风的引导：贯穿内庭上升

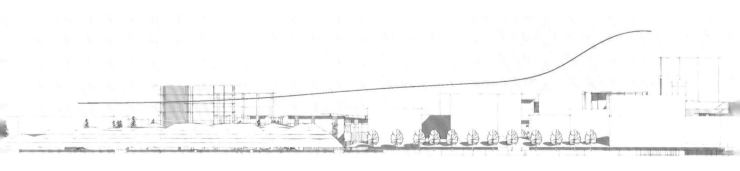

江风的引导：南北立面

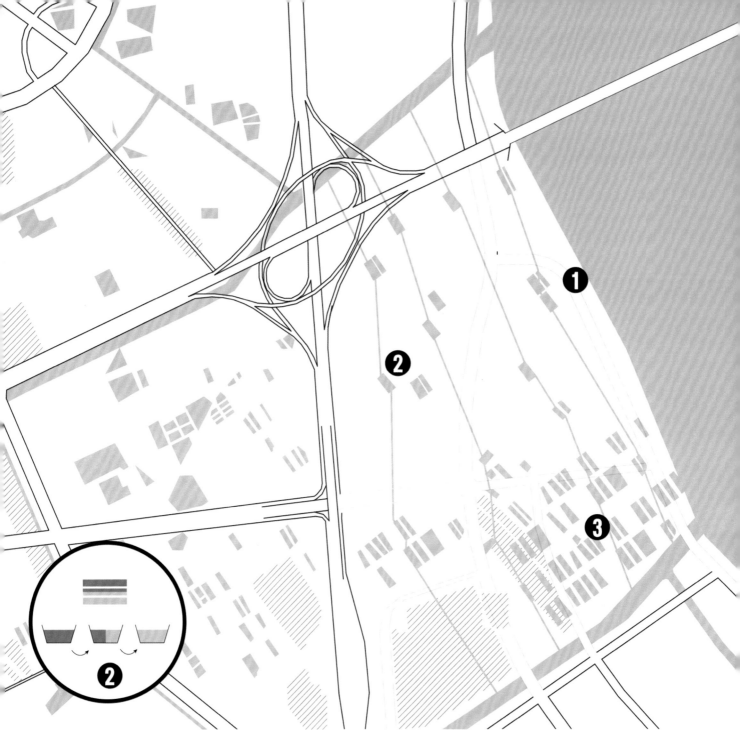

自然水系的重新组织

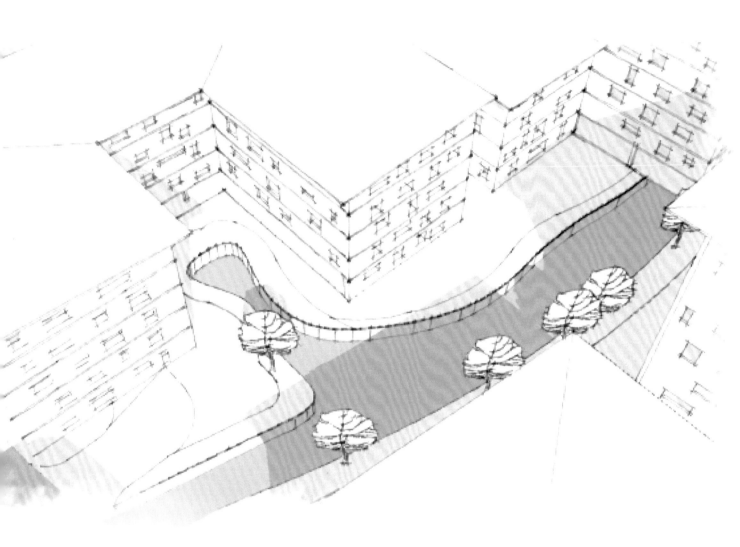

自然水系的引导利用

场地二

运河地块

课程简介

课程目的：本课题关注当今城市化进程背景中的城市集合居住的乌托邦形态。希望学生对于居住类型、城市形态、发展模式等概念有深入了解。

本实验课程试图打破建筑教学不同课程之间的障碍，从地理、历史、伦理、出行、环保、技术、政策等各个不同角度来帮助学生理解居住的意义、了解城市形态和生活模式之间的相互刺激的关系，从而培养学生的全面思考问题的能力。

设计背景：中国近三十年的城市化进程可能是全球有史以来规模最为宏大、最为迅速的一个过程。推倒重建的过程在几乎所有中国的城市中以惊人的速度重复着。而中国城市的天际线在短短的十几年甚或是几年中，可以发生翻天覆地的变化。对于这样的变化我们找不到比"大跃进"这样更贴切的字眼来形容。

伴随着这个前所未有的城市化进程的是：城市肌理的消解和同质化；自然环境的消失和恶化；社会阶层的分裂日益明显等等。面对这些问题，那些尚背负着一些历史遗存的城市碎片，是否有可能在延续记忆传承的基础之上，同时建立新的有效的城市生活模式或是建筑类型？这是本课题所关注的焦点。

课程时间：2013年春季

场地介绍

三堡社区：
这里十年之前是杭州郊区的乡村地区，现在是杭州东北面迅速发展的城市河运中心。该区域西侧紧贴着世界闻名的京杭大运河的支流，向东有杭甬高速穿过，向北是杭州新建的火车东站。

Sanbao District
This site, which is used to be rural area located at suburban Hangzhou 10 years ago, now is planned as municipal sub-center of water transportation. A branch of well known Grand Canal runs along the west side of the site, where Hangyong Highway is not far from the east and the newly built Hangzhou East Train Station is just blocks away from the north.

阶段一地块：26,500 m²　　Phase 1 Site Area：26,500 m²
阶段二地块：86,000 m²　　Phase 2 Site Area：86,000 m²

两块场地 运河地块

地块功能分析

地块现状照片

地块交通分析

100 · 两块场地 运河地块

运河边历史场景　　　　　　　　运河边当代场景

山水城市
LANDSCAPE CITY

全程中方组员：王建辉、窦研、边如晨、周轶玲
中期法方组员：Francois-Luc Giraldeau, Louise Reboul, Fabian Vandevondele

如同社会之伦理探讨人与人之间的行为规范，当审视建筑伦理之时，即是讨论建筑与人、建筑与建筑、建筑与自然的一套行为规范。在承载传统文化和地理环境的运河沿岸，如何建构这块场地的建筑伦理？设计回应是引入传统山水园林的概念，重新构造地势以营造山水城市的意象，复现传统的人居理想，以慢生活与慢建造的方式来实验一种逆城市化与自然建筑的理念。

地势的重建：结合车库所需空间，重建山的意象。
路径的穿插：在场地中穿插一系列的空间流线，形成与路径相连的不同景观与地形，形成类似园林一般的建筑群体。

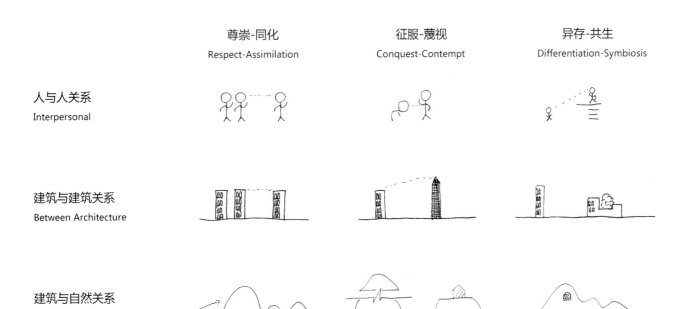

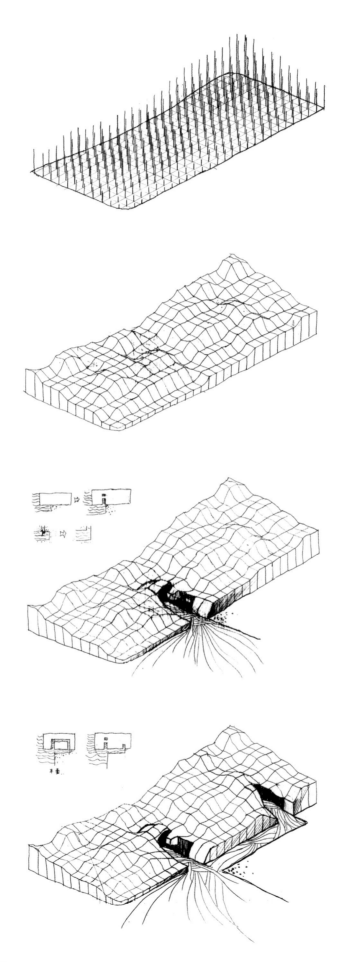

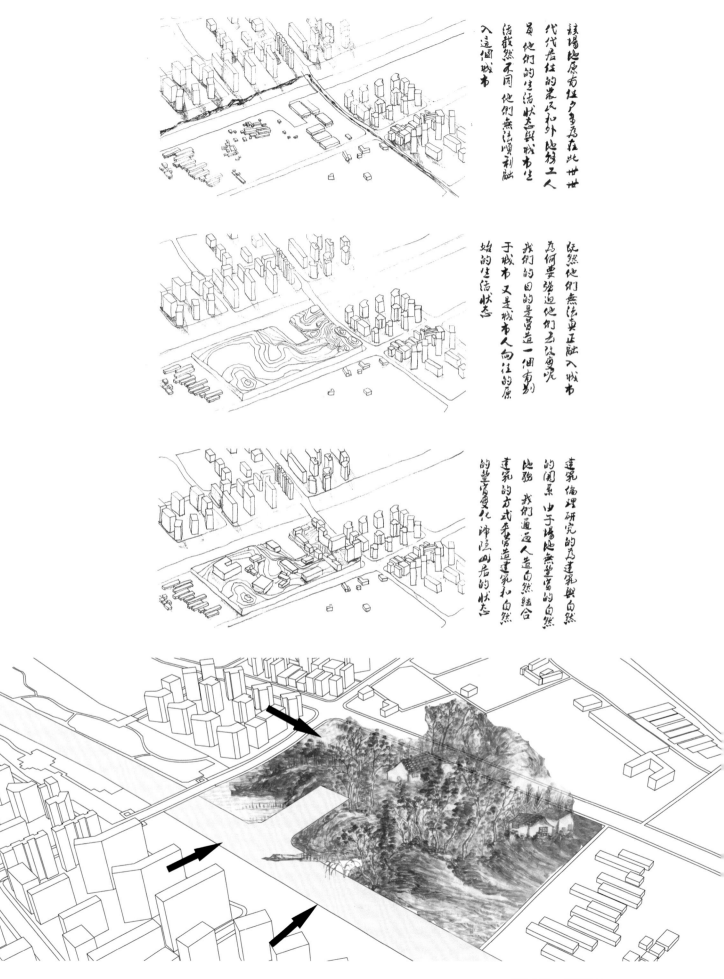

两块场地 运河地块

山层总平面图 1:300

建筑与自然的交织主要体现在
由电梯井到达住宅的过程上
建筑一层架空与地形的关系上
各个平台由廊道串联的路径上
路径中间过渡转换的休闲平台上
有远观点组成了丰富的建筑和自
然的交织图景

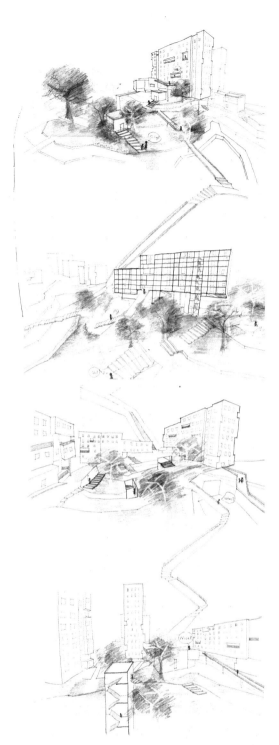
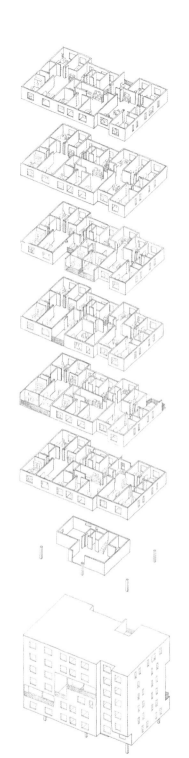
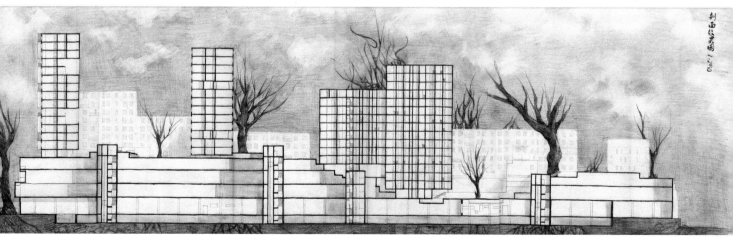

两块场地 运河地块

庭院相生
GARDENS CIRCULATED

全程中方组员：孟婕、袁汉瑜、吴安琪、杜雷
中期法方组员：Francois Julla, Christina Chouiti, Luce Gondran

邻里关系代表着是对于人与人之间的关系的理解，是一种生活状态的表达，也代表着区域与区域之间的关系。邻里交流的缺失是当下城市居住的一大缺陷。如何在集合居住中创造有效的公共空间以催生邻里交流是设计的触发点。

庭院的成型：住宅楼根据日照以基本等距的方式排列，但通过转折形成一个个半围合的庭院。

庭院的相生：单个封闭的庭院缺乏生命力，设计中通过置入一条变化的流线打破一个个封闭的庭院，将它们串联起来。流线自身的空间变化和庭院结合起来形成该庭院独特的空间效果，邻里之间的交流则随着流线在不同的庭院发生。

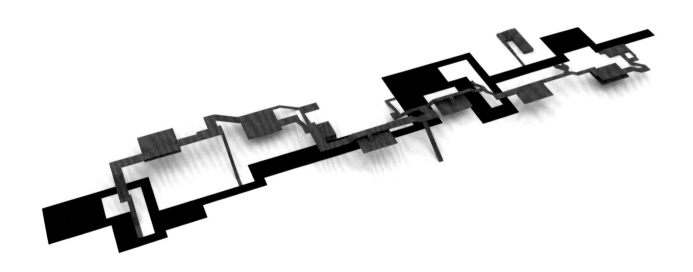

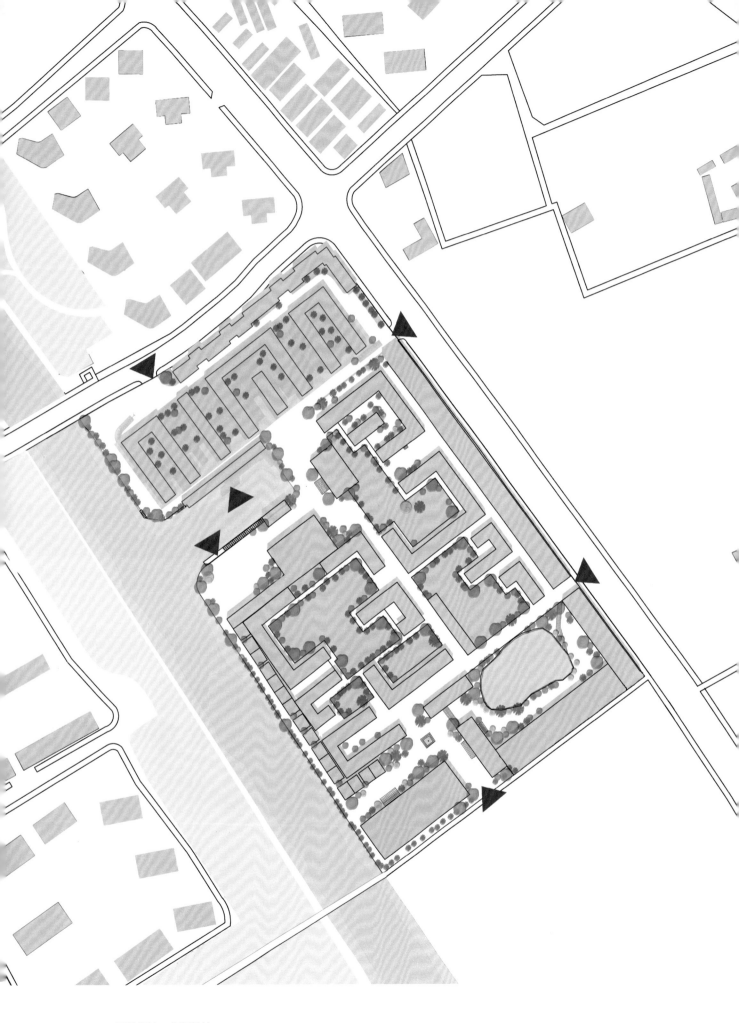

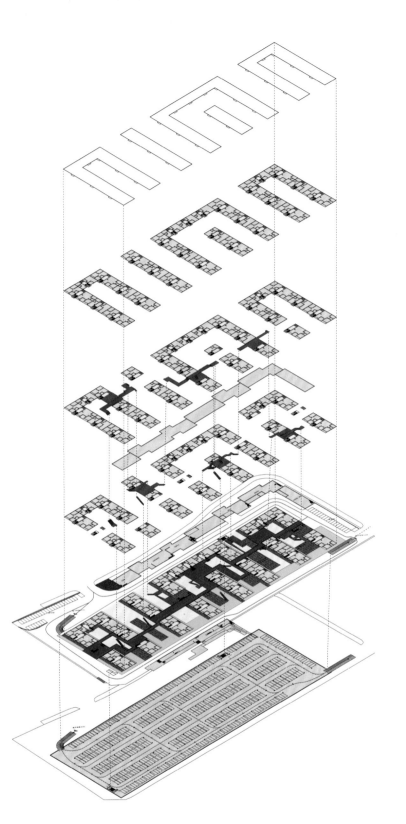

一期地块轴测分解

庭院空间序列

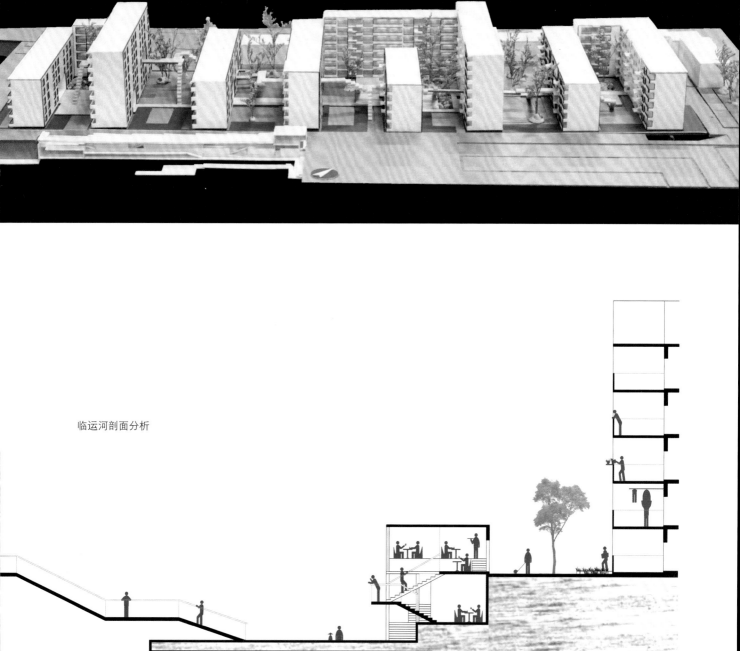

临运河剖面分析

庭院空间序列

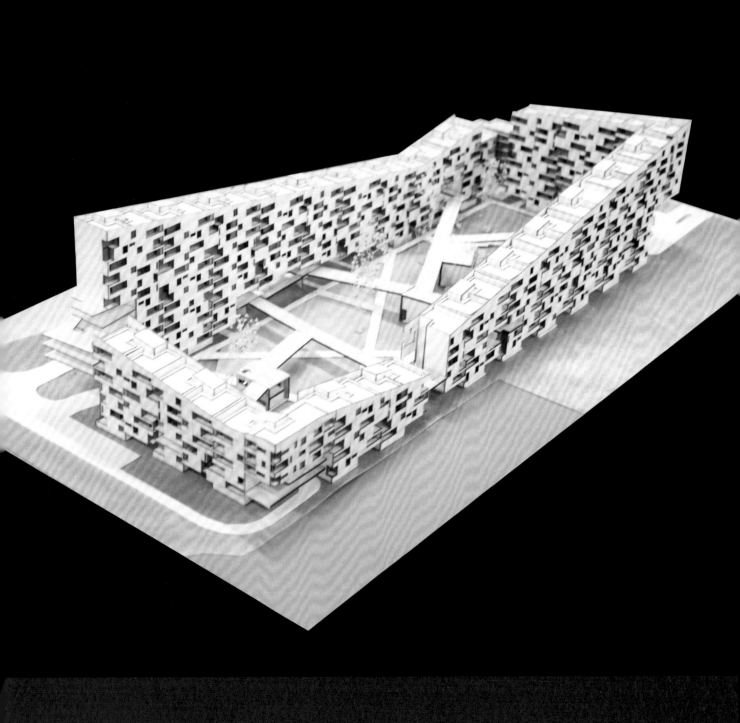

垂直村落
Vertical Village

全程中方组员：丁思圆、徐雅妮、张恺睿、金樊
中期法方组员：Roxane Gentil, Amandine Roy Santoni, Yuan Yuan Wang

城市化的进程抹去了这片土地上曾经的一个个村落，替代的是一个个如同一个模子里出来的住宅单元。在这片原先是乡村的城市地块，是否可以创造垂直的村落，一个三向度的小区，可以把个人自由、多样性、弹性、与邻里生活带回？是否可以提高这个都市村落的密度，以便为住宅创造更多的露天和屋顶花园？

将原有的高度相同的楼层相错半层，呈阶梯状排列在两户间接邻的位置设置露台，将村落的组织垂直化进行，通过廊道和阶梯组织建筑，保持其自发生成的公共空间，将集合住宅与村落实现融合。

在高度变化中形成私密性与公共性的转变，形成不同程度上的活动与事件，也暗示着村落生活中常见的交流模式－一种灰色的交流空间。在设计中将这种交流方式带入城市高层住宅，在高度私密性的存在中介入公共性，形成垂直村落。

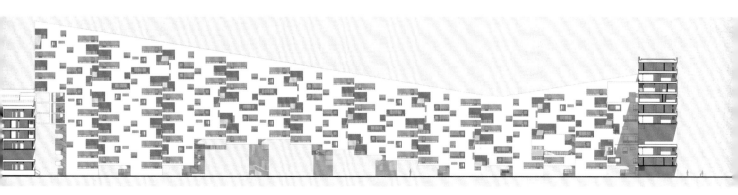

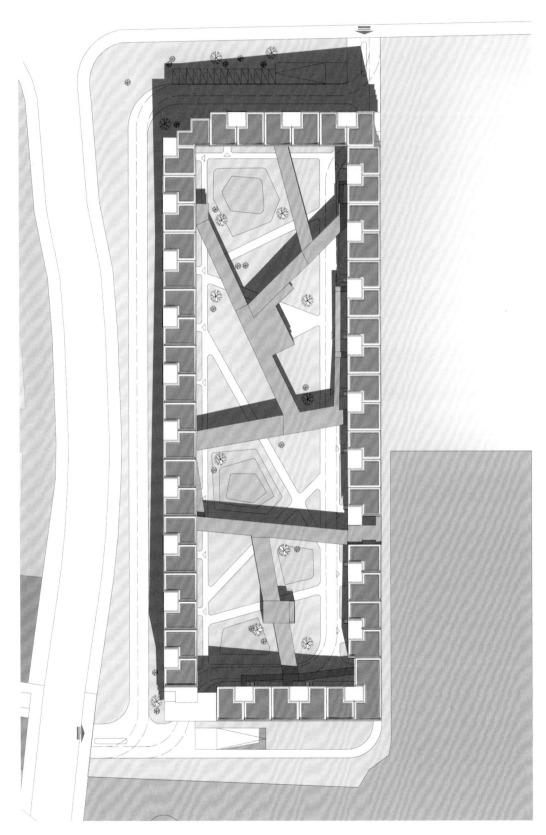

一期地块总平面

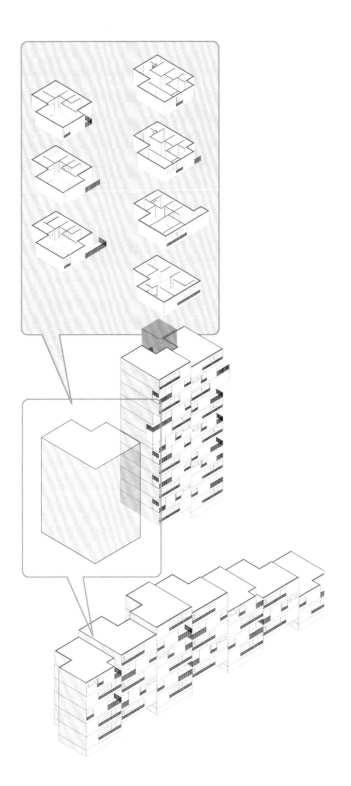
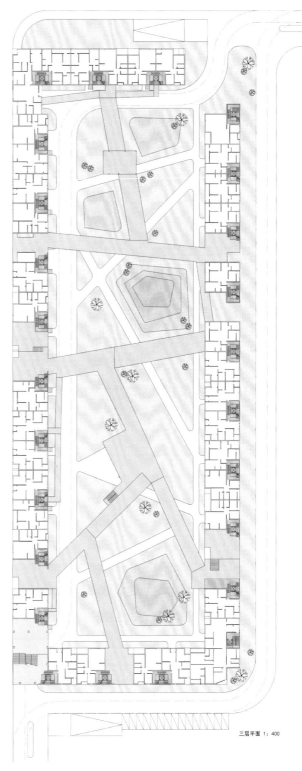

三层平面 1:400

两块场地 运河地块 · 119

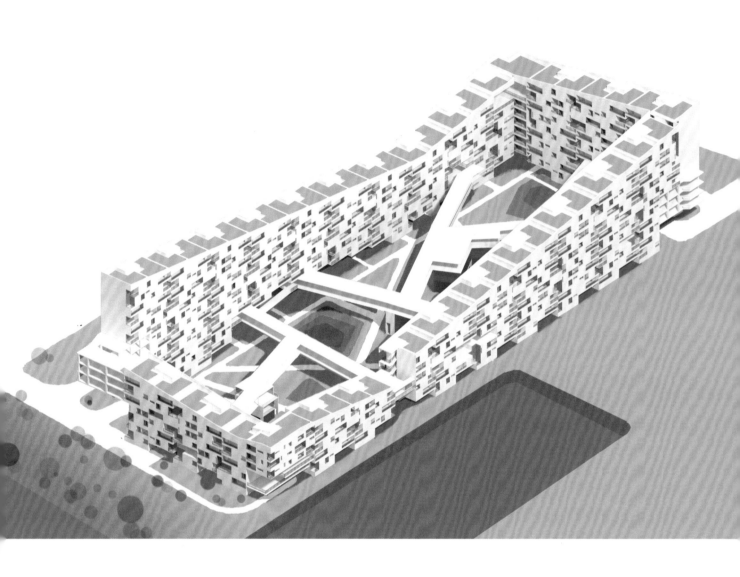

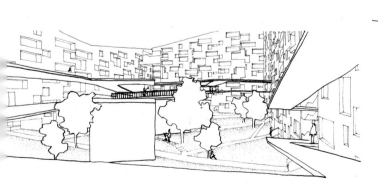

一期地块内院速写

二期地块发展策略

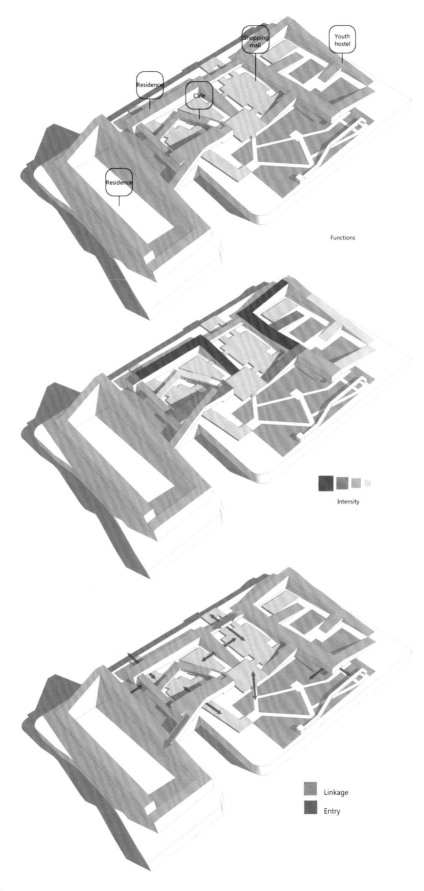

两块场地 运河地块 · 121

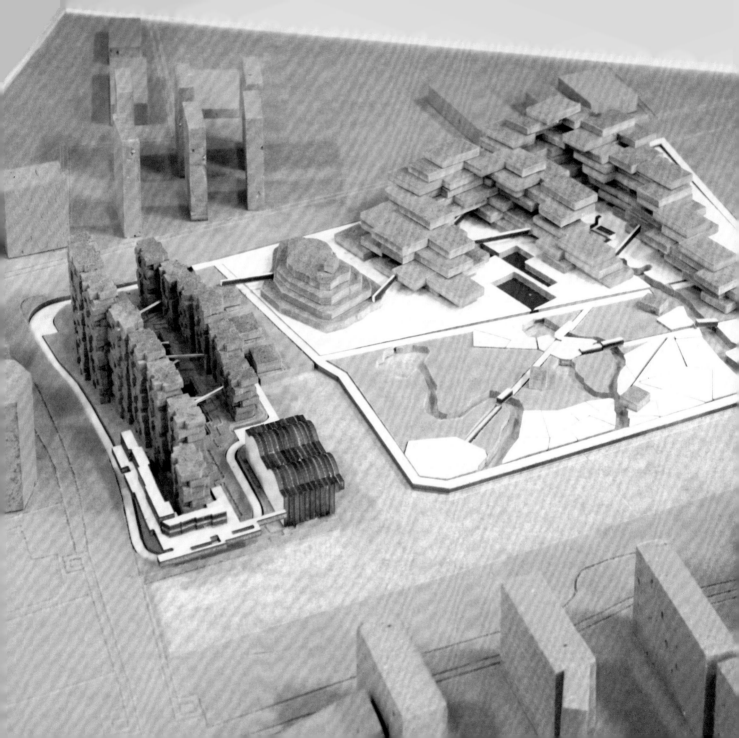

循环代谢
Metabolism

全程中方组员：毛小悲、彭世杰、朱韵婷、马成龙、吴瑶丹
中期法方组员：Philippe Lafleche, David Ottilik, Perle Van de Wyngaert

运河是古代大宗物品运输循环的重要通道，紧邻运河的场地自然也是物质交换和循环的自然发生之地。即使在当下，仍有经营废旧物资的人占据着这块场地。

居住地块上同时置入居住、农业、轻工业、市场等功能，尝试建构物质、资金、能量在这块场地上的代谢循环。期待随着时间的推移，逐渐平衡下来的代谢循环能催生一个在农业、商业、垃圾处理以及居住环境方面都有独特处理方式的、具有一定自治属性的自治小社区。

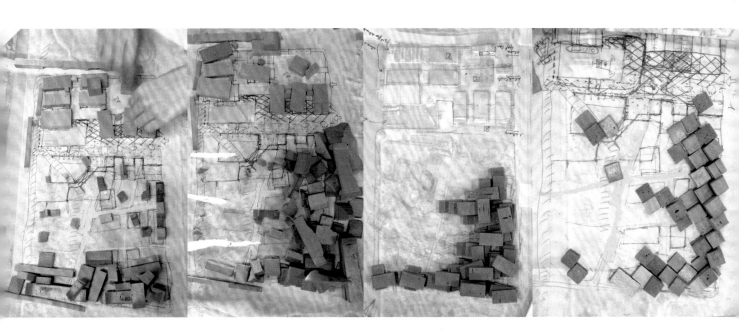

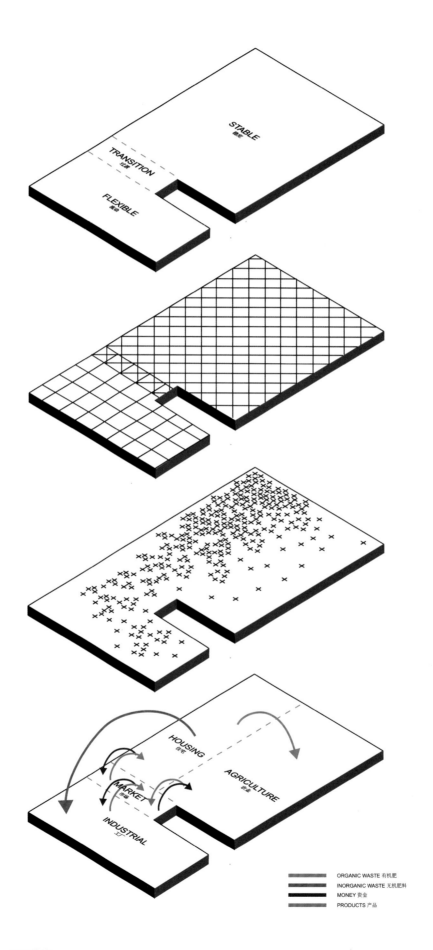

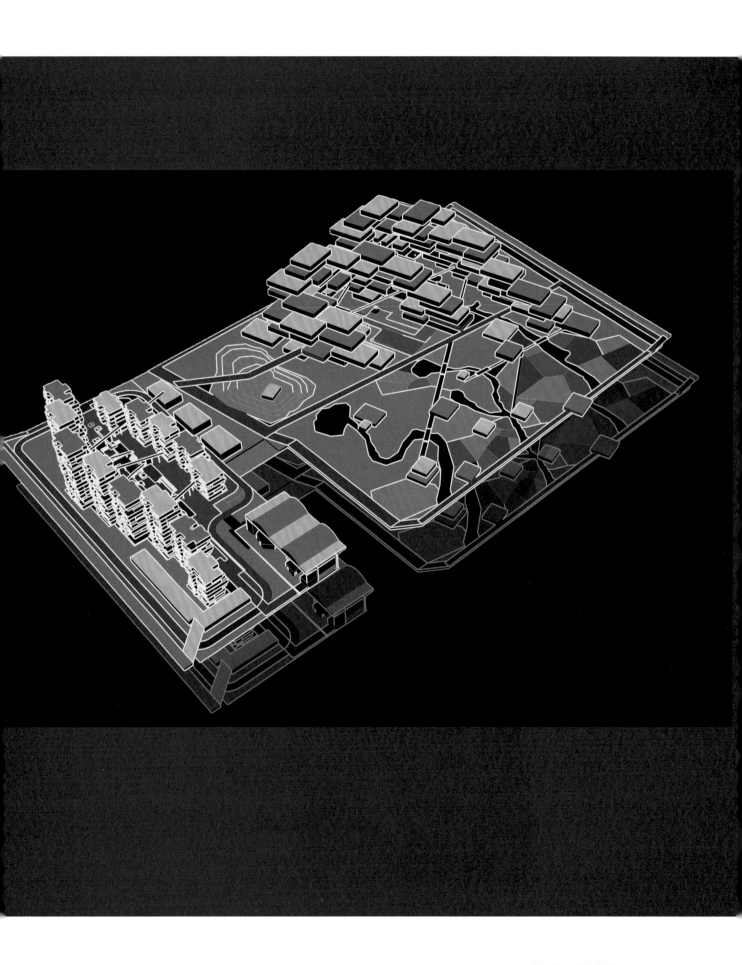

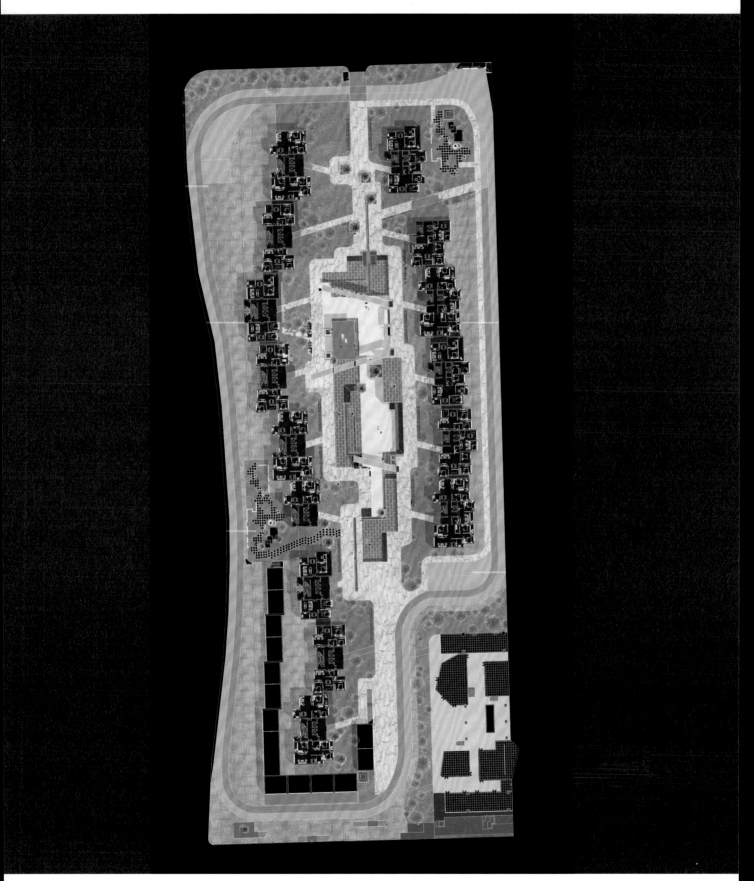

一期地块总平面

126 · 两块场地 运河地块

Elevation

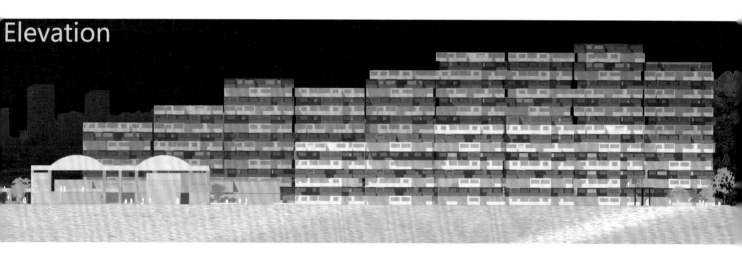

Section

智者乐水

"THE WISE ENJOY THE WATER"

全程中方组员：郭倩、蒋佩、黄佳强、唐慧、黄迪
中期法方组员：Marion Mouny, Roxane Halliez, Lorenzo Giardina

紧邻大运河，水自然成为场地建造的根本与场地景观的元素。尝试利用新型的水资源处理方式：净化、娱乐、农业、能量循环与自然景观，来实现整块场地的自然问题，实现生态建造的设想。

水的净化：大运河的水非自然之水，水质较差。尝试沟通运河与场地内部水系统的联系，设立沉水池进行水资源的交互，实现水体的净化，也达到了自然植物与景观被保留的作用，形成类似于湿地系统的水循环生态体系。

水的景观：在建筑中实现水体与建筑相互呼应的效果，形成水体景观，利用水来实现众多功能体现，是作为该场地独有的生态运转方式，以打破城市均质化这一共性。

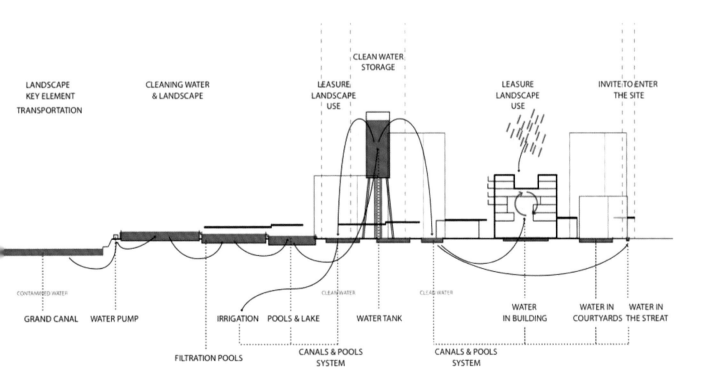

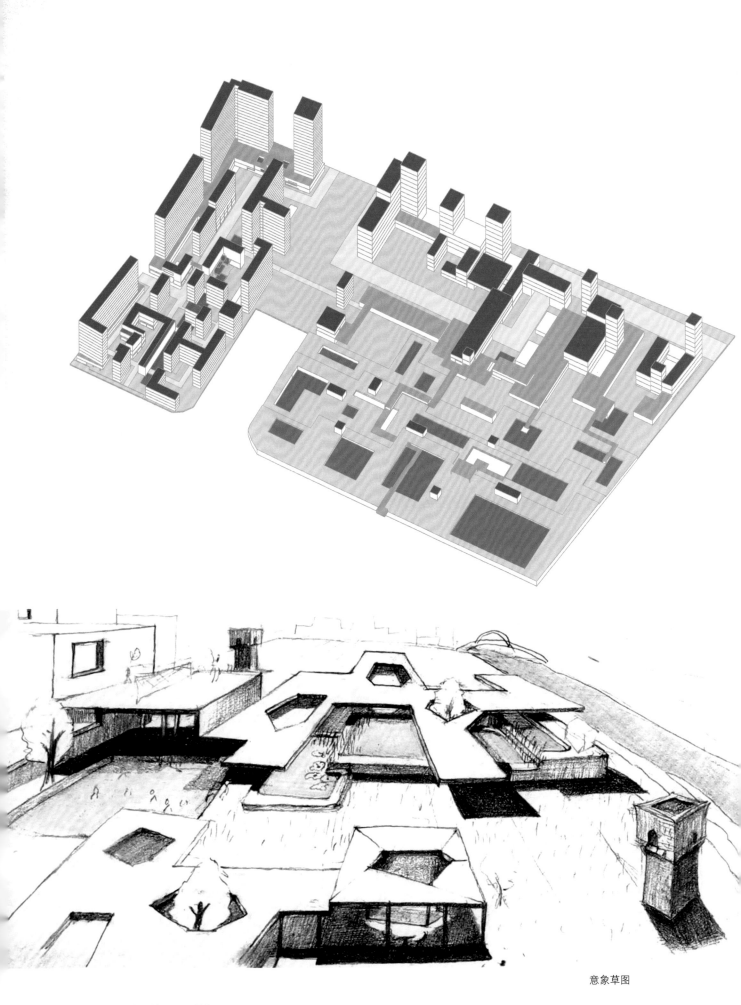

意象草图

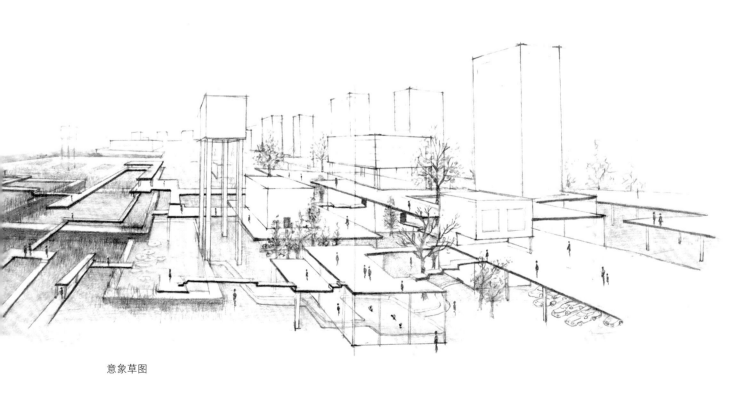

意象草图

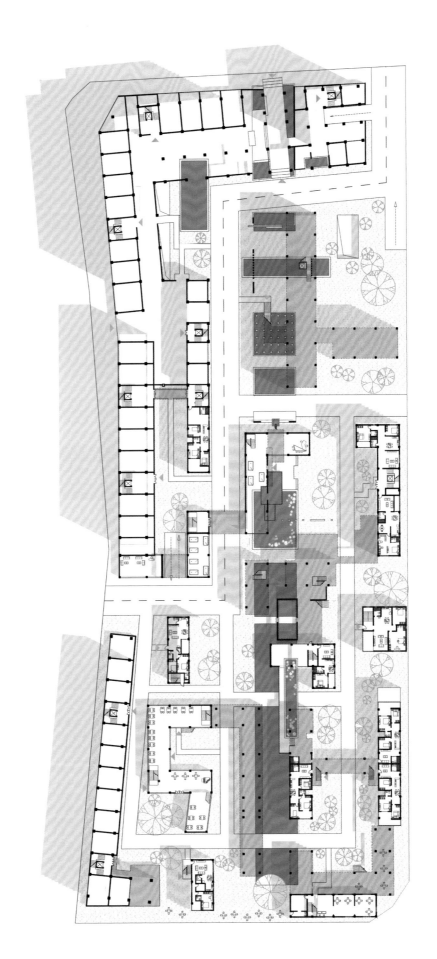

一期地块总平面

132 · 两块场地 运河地块

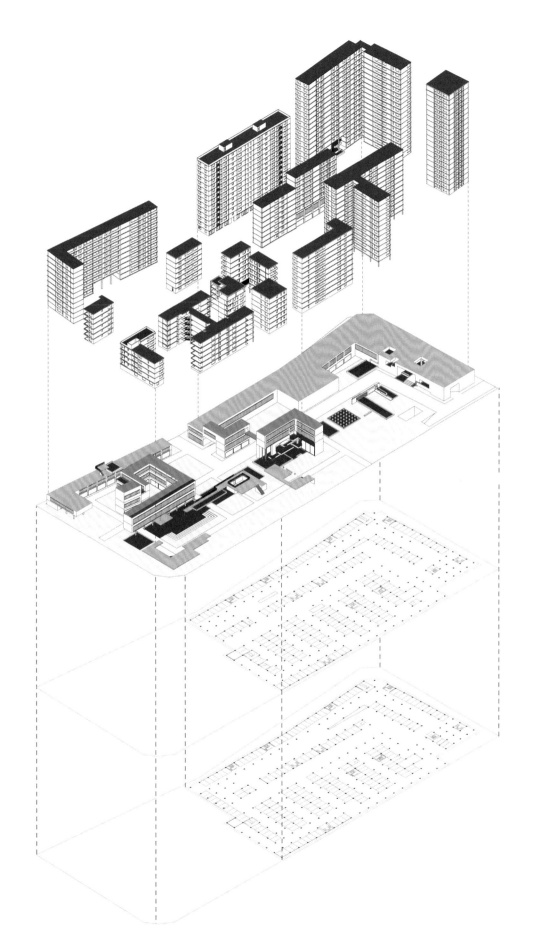

两块场地 运河地块

本书出版受到浙江省高等教育教学改革项目课题的部分资助,在此表示感谢。
课题名称:城市设计专业建设与浙江省新型城镇化建设的协调发展研究;课题编号:jg2015019。

责任编辑：徐新红
责任校对：朱　奇
责任印制：毛　翠

图书在版编目（CIP）数据

城市住区 / 卓旻著. -- 杭州 ：中国美术学院出版社，2017.4
 ISBN 978-7-5503-1320-0

Ⅰ．①城… Ⅱ．①卓… Ⅲ．①居住区－城市规划－设计 Ⅳ．①TU984.12

中国版本图书馆CIP数据核字(2017)第060569号

城市住区
卓旻　著

出 品 人：祝平凡
出版发行：中国美术学院出版社
地　　址：中国•杭州市南山路218号／邮政编码：310002
网　　址：http://www.caapress.com
经　　销：全国新华书店
制　　版：杭州海洋电脑制版印刷有限公司
印　　刷：浙江省邮电印刷股份有限公司
版　　次：2017年4月第1版
印　　次：2017年4月第1次印刷
印　　张：9
开　　本：889mm×1194mm　1/16
字　　数：80千
图　　数：80幅
印　　数：0001-2000
书　　号：ISBN 978-7-5503-1320-0
定　　价：85.00元

版权所有　翻印必究